W9-AUC-799

THE TECHNOLOGY TRANSFER SYSTEM

THE TECHNOLOGY TRANSFER SYSTEM

INVENTIONS

Marketing • Licensing • Patenting
Setting • Practice • Management • Policy

Albert E. Muir

Latham Book Publishing
Latham, New York
1997

Library of Congress Catalog Card Number 97-93356

ISBN 0-9657597-6-8

Printed in the United States of America

I dedicate this work to

my wife Mary Ellen, daughter Justine, and son Basil,

for their enduring encouragement and support.

Contents

Part II — Practice

Part III — Oversight

Appendices

Acknowledgments

My sincere thanks to the individuals involved in technology transfer, at The Research Foundation of State University of New York, and at the State University of New York campuses; to the many inventors, research scientists and faculty who I served through the Foundation's Technology Transfer Office. Thanks to the people in the industrial sector, my contacts in new technology marketing and licensing, and the patent attorneys with whom I dealt in the patenting process. These parties contributed immensely to my experience and ability to undertake this work.

My particular thanks to the Foundation for the opportunity of the experience. The Foundation has been most gracious in allowing me to use corporate policy documents and in-house sample technology transfer agreements for this work, and for this I am greatly appreciative.

However, these parties are not to be blamed for any errors the reader my find. For these I take sole responsibility. I also take responsibility for my use of the Foundation's corporate policy documents and sample technology transfer agreements, which make this book so much more valuable as a reference.

Introduction

The *Technology Transfer System* is comprised of the individuals, institutions, practices, laws and policies involved in the marketing, licensing and patenting of inventions. This book explains how the system operates in the United States today: its historical foundations; the protection it offers under the patent and trade secret laws, and license agreements; the procedures and practices by which transfer occurs; the people and entities that make it happen.

Technology Transfer refers to the conveyance of inventions from one entity to another under license agreements, for the purpose of commercialization. *Inventions* are new technologies in general; ideas for new products and processes, including computer software, that might be entitled to either trade secret or patent protection. The marketing and licensing described in this book focuses primarily on transactions between nonbusiness entities (i.e., independent inventors, universities and colleges, nonprofit organizations, and government) as owners of inventions and commercial enterprises as licensees. Noncommercial owners transfer their inventions to businesses so they may be commercialized.

Except in the case of inventions of independent inventors who may form new businesses to exploit their new technologies, nonbusiness owners generally rely on licensing or outright sale of their inventions to commercial entities as a route to market entry for their ideas. Locating an industrial sponsor and achieving commercialization for inventions, while protecting property rights to new technologies is a core process of the technology transfer system.

The hope of reward spurs invention. In the technology transfer system reward comes primarily when inventions realize their market potential as new products or processes. The cooperating parties are brought together by a common interest in financial reward. Direct beneficiaries include the individuals who originate innovations, institutions that support inventors, and businesses that license inventions in order to commercialize them.

The owners of inventions are rewarded with royalties. Typically, institutional owners share proceeds with their employee inventors. Reward may also include support for patents and additional research funding from an industrial sponsor, for continuing work on the technology. Businesses that license and develop the inventions into marketable products realize their rewards in profits from protected markets, competitive advantage, and technological superiority. For the nation as a whole, technology transfer contributes to technological progress, economic development, and global competitiveness.

Who Could Use This Book

Independent inventors who want to locate industrial sponsors on their own will find this book a valuable resource on all aspects of marketing, licensing, and patenting. Employed researchers, scientists, and engineers will find helpful answers to questions about their rights to inventions and the responsibilities of the institutions that employ them.

The book's comprehensive treatment of technology transfer makes it a useful reference for practitioners within the system who manage invention portfolios; professionals responsible for administering research programs and institutions; business strategists involved in licensing inventions; and public policy analysts concerned with the practical aspects of technology transfer. In addition, the book provides basic material for an introductory course on technology transfer, and would be especially useful as a supplemental text in such a course.

How This Book is Organized

The book contains three parts, first describing the setting and then the practice of technology transfer. This arrangement gives the reader background on technology transfer before discussing specific facts of the system such as transactions between the parties, the enabling laws and practices, and aspects of management and public policy.

Part I. Covering the setting for technology transfer, Part I provides a description of the environment in which the system functions. This section provides a historical context for the present day system by outlining developments such as the origins of trade secret law and patents in early industrial policies. It notes significant inventions and their use in producing increasingly advanced products and processes. The historical perspective also includes important contributors to technological development. Part I concludes with a description of the individuals and organizations involved in technology transfer, and current laws relating to trade secrets, patents, and antitrust.

Part II. Part II addresses the practical aspects of technology transfer. It begins with the characteristics of companies that license inventions and covers the attributes of good licensees. It then describes techniques for locating industrial sponsors. Subsequent sections present the development of a license agreement, the negotiation process, and ways to arrive at reasonable royalties. Part II also provides a mathematical model that takes into account the varied considerations in license negotiations, and concludes with an explanation of the patent application and prosecution process.

Part III. This section addresses the oversight of technology transfer; that is, its management and policy. Management relates to the handling of multiple inventions by the technology transfer offices created by universities and other nonprofit institutions, including government laboratories. A description of one such office, the Technology Transfer Office of The Research Foundation of State University of New York, illustrates the functions and responsibilities of a typical office.

The second part of this section develops a methodology for evaluating the performance of technology transfer offices. The last chapter reviews national technology transfer policy, including the incentives and role of government in the process, with implications for licensors and licensees of new technologies. The book closes with a note on national indicators of technological development.

Note to Readers

An important determinant of successful licensing and patenting is one's skills in the practices of technology transfer. By providing a one-stop reference on the system and important supporting literature, this book offers a means for improving such skills. The purpose of the book is to educate, and to serve as a guide. It is not offered as a legal or financial service. Because of the complexity of the subject matter, the author strongly urges the reader to also consult with a competent authority if professional assistance with technology transfer is required.

PART I

Setting

Chapter 1

Historical Influences

The importance of innovation for industrial and technological progress is well documented in history. A natural consequence has been policy to encourage invention, the source of innovation. Laws have been enacted to protect the rights of owners of inventions, so that they may benefit from their discoveries. Benefits accrue from the commercial exploitation of the new technology, which such rights induce.

The United States Constitution mandates that inventors be granted exclusive rights to their inventions. This positive disposition of the Founding Fathers toward technological progress has become a unique characteristic of the American way. Among the defining historical experiences are those associated with individual rights, the monopoly grant as a policy instrument; inventions, patents, patent statutes, and their importance for industry. Licenses, the legal instrument of technology transfer also appear in early European history. Each of these experiences was translated into practice during the colonial period, and has become a dynamic component of the United States technology transfer system.

HISTORICAL DEVELOPMENTS IN INTELLECTUAL PROPERTY RIGHTS

Trade secret law and patent law evolved separately. Early writers on patent law appear to indicate that a trade secret was not recognized as a property right, as evident in the following comment: "He is not required to make and publish his specification. But then it [the invention] would not be a property; it would, after all be only his secret."[1] Today the holder of a trade secret has the right but is under no obligation to apply for a patent.

Trade Secret Law

Trade secret law has roots in English common law. The latter evolved over time as people came to realize that certain wrongdoings required collective sanction. Wrongdoings were perceived as violations of custom or belief. The first judges were guided by their understanding of the prevailing moral and social codes. Over time, when several judges arrived at the same decision regarding a certain wrongdoing, that decision acquired the force of law. Certain rights accrued to the people as judicial decisions became common law.

The trade secret holder became entitled to the protection of English common law if someone else acquired the secret by means which were offensive to the prevailing moral codes. Present-day legal recourse for the offended party has come to be based largely on the occurrence of a violation of confidence. A violation of confidence is against our business ethics and offends our current morality.

Given the inventor's option to obtain a patent, English courts at first were reluctant to define a trade secret as a protectable interest. This reluctance was overcome by a case heard by the English chancery court in 1851. The case came to influence developments in the United States as well.

The case involved Morrison and Moat, partners in a business using a secret recipe for medicines. They were sons of the founders of the business. When Moat left to start his own business, based on the secret recipe, Morrison brought suit, seeking a restraining injunction. The case was finally settled in Morrison's favor, on the basis that Moat was aware of the trade-secret nature of the recipe and that he was under an obligation to honor that secret. However, while Moat was prohibited from using the secret, the ruling did not address his freedom to divulge it to others.[2]

Early Monopolies and Patents

Patent law on the other hand grew out of early industrial public policy. Its beginnings are founded on exclusivity privileges. Use of the monopoly grant, the basis of present-day patent systems, dates back thousands of years. Aristotle mentions it in his *Politics* as meaning an exclusive sale, and he notes that it was a method used by cities to raise money.

Monopolies existed in Egypt before its occupation by the Romans.[3] The latter used the monopoly grant, particularly in the territories they conquered. However, monopoly policies at that time had no industrial development objective. They were purely schemes to raise money, usually to finance wars.

Support for industrial progress and for the means of achieving such progress is generally traceable to policies in medieval Western Europe. Here, the principle of exclusivity in trade and manufacture assumed the form of monopolies, importation franchises, bounties, licenses, and exemptions. Monopolies and importation franchises were particularly significant for the grants of patents to come.

Monopolies were grants of privileges involving enterprise rights previously available to the population. Importation franchises also conveyed monopoly rights. They were exclusive rights and inducements to encourage foreigners to bring their expertise and technology into the country. This arrangement would produce benefits that were not previously available to the local population. Patents of inventions sprung from this latter policy in that they involved exclusivity, as well as benefits that the population had not previously enjoyed.

In England, privileges in the form of licenses and importation franchises had as their objective the improvement of the living conditions of the people by converting the country's production base from one of raw materials to one of manufactured goods, for domestic consumption and export. Under the franchise, the foreigner was granted protection and exclusive rights in the practice of the desired technology in England.

Licenses permitted foreign workers to be brought to England, to apply their skills. Such privileges were given as royal favors and were not regarded as rights. Today, a license is the instrument under which the legal owner of an invention grants a company the privilege to apply an invention for the purpose of commercial gain. It is a prerogative of the owner to make the grant, absent which the company has no legal claim to the technology.

An early example of an exclusive privilege given to a foreigner for the purpose of creating and fostering domestic industry is the Letters of Protection granted in England in 1331 to John Kempe, a Fleming. The express purpose of these Letters was to encourage the textile industry by having Kempe instruct the English in the "mysteries of the weavers, fullers and dyers" of textiles. This inducement was subsequently confirmed in a 1337 statute that afforded general protection for the domestic manufacture of textiles by prohibiting both imports of cotton and exports of wool. Also, arrangements were made to ensure standards of quality in the products of trade, thereby protecting the consumer.[4]

The instruction afforded under Kempe's Letters of Protection illustrates the willingness of the English people to submit to instruction: an important aspect of the learning, adoption of methods, and practice of skills that make technological progress possible. In this manner, the craftsmen of the time gradually progressed in technological ability and became increasingly capable of creating technology advancing inventions of their own. Letters of protection eventually came to be replaced by Letters Patent, the term we use today, with its monopoly privileges.

Early European Patent Systems

The first known patent of invention was awarded in 1421 to a Florentine architect, Filippo Brunelleschi, for a new kind of ship. This patent embodied the basic principles of future patent statutes. It recognized that the inventor's rights to creative works were inalienable and that the disclosure and practice of inventions was beneficial to society and thereby worthy of inducement. Accordingly, Brunelleschi was granted legal protection in exchange for disclosure and use of his invention.

As opposed to the grants of monopoly privileges noted above, the Italian patent was unique in its recognition of both the invention and the inventor's rights. However, this progressive approach was not unusual for Italy. The Renaissance was fostering the growth of individual rights and liberties. Works of creation enjoyed popular recognition, as did their creators. Furthermore, the rights of the individual were being acknowledged in Roman contract law, also under revival at this time.

In 1474, the Italian Republic of Venice established another milestone in patent history when it enacted the first known general patent statute. This law incorporated the basic principles that appeared in Brunelleschi's patent.

1. It offered protection from infringers to inventors for inventions that they revealed.
2. It required novelty, reduction to practice, and functionality.
3. It stated that the protection it offered was designed as an inducement to creativity and disclosure for the benefit of society.
4. It provided an applications procedure.
5. It stipulated a term for the patent,-10 years.
6. It reserved a right to government to use the invention.
7. It granted the patent as a right and not as a royal favor.

Patents for inventions as incentives for technological development and new industry made their appearance in English law more than a century following the Italian enactment. True patents for inventions began in England in the

1560s. The first one may have gone to an Italian, Giacopo Acoutio, for a twenty-year term. Acoutio recommended to the British Crown that it adopt a patent system, citing the system in his mother country as an example. Eventually, the patent he received contained the basic principles of intellectual property protection first cited in the Brunelleschi award.[5]

In England, patents continued to be issued as general grants of monopoly privilege. True patents of inventions constituted only a small fraction of the accumulating awards. Furthermore, abuse of monopoly grants had become widespread, and patents of inventions became tainted by this association. Finally in 1624, England passed a general ban against monopolies in the Statute of Monopolies. However, patents of inventions were exempted.

The granting of patents in England continued to be regarded as a royal favor, as opposed to the Venetian system in which an invention belonged to an inventor as a natural right. Also, the English law did not include an applications procedure and did not prescribe penalties for infringement, both of which were contained in the Italian statute.

EARLY EUROPEAN SCIENTIFIC AND TECHNOLOGICAL DEVELOPMENTS

New ideas came to Europe mainly through Italy and southern Spain. They came from other parts of the world including China, India, and the Arab world. Among the technologically promising occurrences in these other parts of the world is an event in Africa. Hero, an Egyptian living in 120 B.C. demonstrated the principles of the steam engine and jet propulsion, which were to revolutionize industry and travel 2000 years later.[6] The Arabs offered knowledge concerning astronomy, chemistry, and mathematics, with influences from the Greeks. They made important contributions to Europe's command of technology in such diverse fields as mathematics, medicine, optics, chemistry, cosmetics, metallurgy and shipbuilding.

Response to the new knowledge lead to revolutionary changes throughout medieval Europe. Here the early technologists were aggressive in developing applications of the information, advancing their skills in directions that remained forbidden in other cultures of the time. They dissected bodies for medical knowledge. They made paints and dyes, mordants and glass, thereby advancing the practical arts. They sought after the inner contents of the earth in mining, and brought the Age of Discovery with explorations that opened the way to the New World.[7]

Fame and fortune attended those who ventured out to faraway lands. Voyagers returned with goods, information, and strange, exciting stories, stimulating interest and demand for new and better means of travel. Not surprisingly, the first known patent for an invention, the one granted to Brunelleschi in 1421, was for a novel means of navigation, a new ship.

The Age of Discovery gave rise to a class of instrument makers who provided the voyagers with inventions such as compasses, quadrants, sextants, and telescopes. Among the greatest inventions of this era was the weight-driven clock. This invention was significant in raising the level of technology available to the craftsmen, enabling them to try new methods of transforming and applying motion.

For the most part, however, these craftsmen were applying scientific principles that they were not able to explain and that they did not fully comprehend. These principles later formed the basis of fundamental discoveries in science.

Such was the experience of the medieval alchemists in their attempts to convert base metals to gold. Although they failed in their primary objective, they provided important insights into the properties of substances. They learned, for example, that mercury imparts metallic properties to material, that sulfur imparts flammability, and that salt imparts resistance to heat.[8] Also among the important chemicals discovered by the alchemists was sulfuric acid, which is able to decompose substances quickly and which today is among the most important mineral acids used in industry. Alchemy eventually gave rise to the science of chemistry. Robert Boyle, one of the first chemists, published the specific modern criterion of an element in 1661.

The Renaissance contributed the printing press, making possible large-scale production of books, journals, and newspapers and beginning the era of learned written discourse. During the next two centuries, books that contained carefully detailed illustrations of machines and machine components emerged in Italy, France, and Germany. These books, called "theaters of machines," featured military, agricultural, production, and water-raising mechanisms. Among the components were gearing, cams, pistons, and cylinders, mechanisms that comprised the foundations of the Industrial Revolution.[9]

The mechanical principles evident in the inventions of medieval and Renaissance Europe became understood in the seventeenth century and underlay much of the scientific discovery characterizing this period. In particular, Galileo Galilei (1564-1642) laid the foundations of the general science of mechanics and elaborated the concepts of work and energy, which are among the fundamentals of physics and engineering.

Galileo made revolutionary contributions to the understanding of the deployment of power in machines and to theories concerning the strength of materials and structures. He is credited with replacing the crafts, and empirical inventions of the earlier era with scientific technology. Towards the end of the seventeenth century, Guillaume Amonton advanced the use of air as a motive force, thereby laying the groundwork for the later development of the airplane.[10]

Thomas Newcomen developed an atmospheric steam engine as a practical application of scientific theories developed by Galileo and other continental scientists concerning the use of atmospheric pressure and steam to harness the power of "fire." The Newcomen engine was one of the few early "fire-engines" that actually worked, and it competed favorably with the other sources of power at that time, namely wind, water, and muscle. In particular, the Newcomen engine was used successfully to pump water out of deep coal mines.

It was also during the seventeenth century that the term *technology* was coined. This century saw writers begin to define the scientific method and to develop ideas for managing the science and technology enterprise. Prominent among them was Sir Francis Bacon, who authored a methodology called the reformed inductive method. Bacon provided a clear differentiation between science-based and empirical inventions. More important, Bacon was the first person to describe a social and political program for science and technology.[11]

The eighteenth century saw the onset of what is now commonly known as the Industrial Revolution. The spread of machinery accelerated the demand for fuel, and that demand continued to spur the development of power technology. In 1769, James Watt secured his patent for the condensing steam engine and also obtained patents for an atmospheric engine with condenser, a steam engine with condenser, and other possible forms of steam engine. Watt's engine was an improvement on the engine developed earlier by Thomas Newcomen. By exercising his patent rights, Watts became a successful businessman.

Also in 1769, Richard Arkwright obtained a patent for his water-frame for the mechanical spinning of cotton and wool, enhancing the level of technology in the textile industry. Soon after, Joseph M. Jacquard invented the automatic pattern loom, introducing a binary coding technique that has become basic to modern computer technology. Electrical science advanced in this era also with the discovery of electric current and the invention of the electric battery by Alessandro Volta in 1796.

New applications of power technology gave rise to its uses for transportation. In 1804, Richard Trevithick made the first steam locomotive capable of drawing trucks on rails. The precision and standardization required to produce large numbers of engines contributed to the growth of another industry, the industrial machine tool industry.[12] This industry in turn facilitated the growth of the textile industry and the birth of the factory system.

Fire hazards resulting from the use of wooden structures for factory buildings spurred technical progress in the iron industry, and again Galileo's theories were used, this time concerning the strength of materials capable of constructing multi-story fireproof buildings able to withstand the vibrations of factory machines in operation.

The science of chemistry experienced another major breakthrough during this period. Up until then it was thought that organic compounds could not be derived from minerals. Chemists believed that only animals and plants could produce organic substances by use of a mysterious energy that they possessed. This belief was known as vitalism. Organic substances could be transformed through extraction into other products such as drugs, dyes, tea, and coffee, but they could not be synthesized.

Then by chance in 1828, the German chemist Friedrick Wohler synthesized urea while working with cyanogen and ammonia. However, it was argued that vitalism was still at work, since the source of these compounds was dead animal bones. A student of Wohler's, Adolf Kolbe, finally put an end to this belief when he synthesized acetic acid in 1844 wholly from inorganic substances.[13] This breakthrough lead to major advances in such areas as drug synthesis, plastics, and synthetic dyes and paved the way for the revolutions in molecular biology and biochemistry that we see today.

In France particularly, great emphasis was being placed on public policy. France created institutions such as the Ecole Polytechnique, which offered training in engineering and the physical sciences and offered prizes to the most talented. On a per capita basis, France had the most scientists. It laid the theoretical basis for significant technological applications used in Britain and published textbooks considered superior to those of the English.[14] England excelled in the application of science and technology to its industrial base.

EARLY AMERICA

In early America, industrial and technological change came from various forms of technology transfer. Immigrants to the New World brought with them knowledge of the scientific, technological, and industrial practices of their motherlands. The early colonists made use of these practices as a means of creating economic self-sufficiency.

English mechanics were brought over by colonial business and political leaders to set up mechanized production. As the new technologies of the Industrial Revolution came into practice, the best of them were soon adopted in the colonies. One example is the Newcomen engine. Immigration of technical people was also an important factor in America's industrial and technological development, a celebrated case being Samuel Slater's role in building the first successful textile mill.[15]

Common law also came to the colonies with the first settlers. Here, it developed separately in the different colonies as case law, as was the practice in England. However, new legislation here was subject to review by authorities in the mother country.[16] The United States trade secret law evolved in this manner from this English common law, with changes over the years due to legislative and court actions. The jurisdiction of the States over trade secret cases applies to this day.

Developments in patent rights was also a prerogative of the individual colonies during the colonial period. The first patent was issued in Massachusetts in 1644 to Samu(el) Winslow for a new method of making salt. The patent was issued for a ten-year term. That same year, the General Court of Massachusetts adopted a "Body of Liberties", with a clause reminiscent of Britain's Statute of Monopolies, prohibiting monopolies but exempting patents of inventions from the prohibition.

The first general patent statute was adopted in South Carolina as a clause in that state's Copyright Law of 1784. It granted a term of fourteen years for patents.[17] The South Carolina general patent statute was different from the Statute of Monopolies in that it explicitly encouraged the arts and the sciences. Similar provisions have been incorporated in the United States Constitution. Unlike the case for trade secrets, with independence came a federal statute establishing patent rights.

INVENTIONS AND THE UNITED STATES CONSTITUTION

Support for technological progress, and an important means by which it is achieved, namely through exclusivity by grant of a monopoly right was an accepted practice in the western world at the time of the settlement of this country. As a result, both trade secret and patent law became established as foundations of intellectual property protection in the new nation. Rights to exclusivity as a means of inducing inventions became a mandate in Article I, Section 8, Clause 8 of the United States Constitution. The Founding Fathers provided in Clause 8 that Congress shall have the power,

> . . . To promote the progress of science and the useful arts by securing for limited times to authors and inventors the exclusive rights to their respective writings and discoveries.

In this way, the ideas, policies and practices that had taken many hundreds of years to evolve in the mother countries of the settlers and immigrants had become the foundation for industrial and technological progress in the United States. The new nation built creatively on this foundation to become the world power it is today. And, intellectual property rights, embodied in patent and trade secret law continue as powerful influences in business competitiveness, and in the country's economic, scientific, and technological advance.

Notes

[1] Louis Orenbuch, "Trade Secret and Patent Laws," *Journal of the Patent Office Society*, Vol. 52, No. 10 (October 1970): 639.

[2] Ibid., 642-643.

[3] Harold J. Fox, *Monopolies and Patents: A Study of the History and Future of the Patent Monopoly* (Toronto: The University of Toronto Press, 1947), 19-22.

[4] Ibid., 43-45.

[5] Bruce W. Bugbee, *Genesis of American Patent and Copyright Law* (Washington, D.C.: Public Affairs Press, 1967), 17-30.

[6] *World Book Encyclopedia*, Vol. 11 (Chicago, London: Field Enterprise Educational Corp., 1974), 691.

[7] D. S. L. Cardwell, *Technology, Science and History* (London: Heinemann, 1972), 4-7.

[8] Issac Asimov, *Asimov's New Guide to Science* (New York: Basic Books, Inc., Publishers, 1984), 264-265.

[9] Brooke Hindle and Steven Lubar, *Engines of Change: The American Industrial Revolution, 1790-1860* (Washington, D.C.: Smithsonian Institution Press, 1986), 10-11.

[10] Cardwell, Technology, 114.

[11] Ibid., 30-89.

[12] Ibid., 100-122.

[13] John R. Holum, *Introduction to Organic and Biological Chemistry* (New York: John Wiley & Sons, Inc., 1969), 2-3.

[14] Cardwell, *Technology*, 123-124.

[15] Hindle, *Engines of Change*, 60-61.

[16] Allen E. Farnsworth, *An Introduction to the Legal System of the United States* (New York: Oceana Publications, 1983), 1-12.

[17] Bugbee, *Genesis*, 60-93.

Chapter 2

The Setting for Technology Transfer in the United States

The technological development of the United States has depended on the friendliness of its institutions, laws and practices, and these have had a decisive impact on economic growth. They have been influential in fostering the scientific way and in determining the direction of progress. A fundamental contributing factor has been the commitment to technology expressed in the founding legislation.

The technological orientation of the nation has also resulted in a fertile environment for technology transfer. Trade secrets and patents have matured as instruments of legal recourse. Science has become integral to invention and has been adopted by business in product and process creation. Also, licensing is practiced as a viable business development alternative. Universities and other nonbusiness entities have grown to be important new sources of invention in this process. Laws have also been enacted prohibiting anti-competitive practices arising from undue market power.

TRADE SECRET AND PATENT LAWS

In the United States today, the owner of an invention is entitled to protection under two bodies of law. Trade secret law and patent law. Trade secret Law forbids breach of faith and the appropriation of another's secrets. The patent law encourages inventions and the arts by granting certain monopoly rights. Trade secret law is based on case law and is founded on English common law. The United States patent system, on the other hand, is based on federal statute. It too has roots abroad.

Trade Secret Law

The American colonies had developed separate legal systems as English subjects, and their prerogative to appeal to common-law doctrine was not altered by the Constitution. Trade secret law evolved in common law under this tradition of separateness. While all federal laws derive from the Constitution, Section 34 of the Judiciary Act of 1789 states that

> the laws of the several states, except where the constitution, treaties, or statutes of the United States shall otherwise require or provide, shall be regarded as rules of decision in trials of common law in cases where they apply.[1]

Development of United States trade secret law was particularly influenced by the decision in the English case *Morrison* v. *Moat* (see Chapter 1). That case stressed the sanctity of a trade secret but also recognized its value as property. The landmark trade secret case in the United States is *Peabody* v. *Norfolk,* decided in Massachusetts in 1868. This case ruled in favor of an inventor, declaring that

> If he invents or discovers, and keeps secret . . . whether a proper subject for a patent or not . . . he has a property right in it which a court of chancery will protect against one in violation of a contract or breach of confidence undertakes to apply it to his own use, or to disclose it to third persons.[2]

An important determinant of rights to a trade secret was its actual use as a commercial product or process. In this event, all people who learned the secret because of their involvement with the commercial enterprise were obligated to maintain the owner's rights to the secret. Thus, a former employee was prohibited from practicing the invention or conveying the secret information to a new employer.

However, if a person outside the commercial venture or other business developed the same invention independently or derived it from a known product or process, the first claimed owner had no legal recourse if a new business was developed as a result. These conditions are familiar in practices today.

This condition of trust applied to a licensee as well. States enforced licenses requiring payment of royalties regardless of whether an invention was patentable. Such contracts continued to be enforced even when the secret became public. In a 1959 case, *Lear* v. *Atkin*, the court ruled that all secret know-how acquired under a license is protectable, but a licensor cannot prevent a licensee from challenging and causing a patent to be declared invalid to avoid paying a royalty.[3]

Patent Law

Shortly after passage of the Constitution and its mandates concerning inventions, Congress passed the Patent Act of 1790. This legislation created a commission composed of the Secretary of State, the Secretary of War, and the Attorney General to examine patent applications. George Washington signed the first Letters Patent, the official document conferring rights to an invention, in that same year. The patent was awarded to Samuel Hopkins for a chemical process that employs ashes from wood to make potash, an ingredient used to make soap.

During the first years of the Patent Act, the commissioners awarded patents through an examination procedure. However, they were unable to process patent applications in a timely fashion. This lead to complaints from the people.

The government responded in 1793 by instituting a registration system. Under the new procedure issuance became basically a clerical procedure, with patents awarded to all applicants upon satisfaction of certain formal requirements.

Under the registration system, the proposed invention no longer needed to be deemed sufficiently important or useful. Conflicting claims to an invention would be resolved in the courts. This system led to confusion and eventually had to be abandoned. A Patent Office, with a sufficient staff of examiners, and the position of Commissioner of Patents was finally created by Congress in the Patent Act of 1836. In that year, a numbering system was instituted that continues to this day.[4]

Through the years, the variety of new ideas that could be patented has been expanded. In 1842, patent rights were extended to designs, and in 1930, patents began to be granted for sexually produced plants other than tuber-propagated plants. Patent-like rights to sexually produced plant varieties were conferred by the Plant Variety Protection Act of 1970, and in 1988, the first patent was issued for an animal: a transgenic mouse.

A more workable definition of what constitutes a patent was developed in 1952, and a new criterion for patentability was added: the requirement of nonobviousness to one skilled in the art. The life of a patent, originally 14 years, was extended in 1861 to 17 years.

Enactment of the Uruguay Round Agreement Act (URAA) on December 8, 1994, once again changed the term of a patent, effective June 8, 1995. The URAA implements the General Agreement on Tariffs and Trade (GATT) and affects members of the World Trade Organization (WTO), which has a membership exceeding 100 countries.

Under the URAA, the life of a patent is now 20 years from the filing date of the patent application, with allowance for extensions of the patent term for up to five years to compensate for delay caused by interference proceedings, secrecy orders, and appellate review.[5] The term of a patent can also be extended for up to five years for delays caused by regulatory review under the Drug Price Competition and Patent Term Restoration Act of 1984.

The URAA affects the provisions of U.S. law relating to infringement and inventive activity and provides for the filing of a provisional patent application. Importation and offers for sale of patented goods will now be considered infringing acts, and invention outside the United States, if provable, will be accepted as evidence in interference cases. No longer is proof of inventive activity or prior art for application purposes restricted to happenings in the United States.

The URAA is a recent example of U.S. collaboration with other countries to protect intellectual property rights. This law is exceptional in its scope in that it affects how long rights last, rights of parties with respect to both issued patents and pending applications, and options with respect to filing. An earlier U.S. collaboration with other nations was the Paris Convention for the Protection of Industrial Property, signed in 1883. More than 70 nations took part in that agreement. This agreement provides that member countries, known as the Paris Union, shall accord the same patent rights to citizens of member countries as to their own.

Under the Paris Convention, if a patent application is preceded by an earlier filing in a member country, the applicant enjoys the benefit of the earlier filing. The earlier filing establishes a priority date for applications. So, if a publication occurs after a U.S. filing, for a period of one year after such filing the applicant is still able to file in other member countries, even though the latter may require absolute novelty as a condition of patentability.

In 1978, the United States became a signatory to the Paris Cooperation Treaty (PCT), under which a patent application in one country is recognized by all member countries. However, this application does not lead to a single international patent. Applicants must later have their applications examined in the member countries designated in the PCT filing. The Budapest Treaty, signed in 1977, establishes requirements concerning the availability of biological material for patenting and patent purposes.

Countries also enter into regional agreements. The North American Free Trade Agreement (NAFTA), signed in 1993, is an example. It addresses mutual concerns of the

United States, Mexico, and Canada. The intellectual property provisions of NAFTA have largely been incorporated into the subsequent GATT legislation noted above.

TECHNOLOGY DEVELOPMENT AND TRANSFER IN THE 19TH CENTURY

The enhancements in patent law accompanied a growing exercise of patent rights. Significant amounts of income was earned. Others achieved success by obtaining rights from the original inventors and forming patent pools.

Thomas Howe is an example of an individual who became wealthy on royalty income. Howe received a patent for a sewing machine in 1846, but he made his fortune from royalties paid by infringers. One of the infringers was Isaac Singer. Singer obtained a patent in 1851 for a sewing machine that was an improvement on the Howe machine. With Howe's consent, Singer combined Howe infringers into a patent pool, the first patent pool in the United States. A successful business resulted, with Howe as the recipient of royalty income. Both became wealthy men.[6]

Another source of new business opportunity to emerge in the 19th century were government contracts. In 1879, the federal government awarded Eli Whitney an army contract to build guns. In fulfilling the contract, Whitney developed the use of standardized interchangeable parts for articles of manufacture. Previous to this, components of articles of manufacture had been made separately. Each such article was a combination of its own uniquely made parts and not an assembly of uniform standardized items as was made possible by the Whitney invention.[7]

The method of mass production was significantly enhanced by the emergence of the industrial machine tool industry, featured at the 1851 Paris Exhibition. At this Exhibition, the United States demonstrated the turret and capstan lathes.[8] The former, known for its association with production of the Colt revolver, enabled a single unskilled operator to mass produce screw-threads.

Americans also made improvements on important imported technologies of the day. The steam engine, with its applications in transportation, power generation for factory machines, and mining, are cases in point. Oliver Evans, an American, improved upon the high-pressure steam engine invented by the Englishman Richard Trevithick. Trevithick's engine used less than 30 pounds of steam pressure. Evans' improvements enabled the engine to use several hundred pounds of pressure.[9]

Thomas Edison improved on the attempts of the day to develop practical electric lighting, resolving the problem with his patent of the incandescent electric lamp. Edison also designed the two-phase generator, thereby solving the problem of electric load associated with large-scale use of lighting and caused when lights were turned on and off. Other important inventions by Edison were improvements to the telegraph and the telephone. In his lifetime, Edison was issued more than a thousand patents.

Edison's research laboratory is said to have set the pattern for present-day commercial laboratories, and the company he founded, the Edison Electric Light Company, combined with others to form the Edison General Electric Company. The Edison General Electric Company later merged with the Thomas-Houston Electric Company, forming the General Electric Company.[10] General Electric continues today as a major producer of electrical products.

TECHNOLOGY DEVELOPMENT AND TRANSFER IN THE 20TH CENTURY

Many technological developments exhibited promise of revolutionary impact at the turn of the century but their benefits were not immediately realized. Notable examples are the automobile industry, air transport, the radio, sound reproduction, and the telephone, all of which are now common features of everyday life.

The 20th century has also seen the development of products that were unknown until recently. And the rapidity of modern development is such that entirely new technologies are themselves undergoing revolutionary change and are already promising more fundamental changes in the ways people live and do business.

The technological growth in this century has been built on continuing applications of a rapidly advancing science and engineering knowledge base and on refinements of techniques. This has been true for products and processes already known by the turn of the century. Applications have also resulted in the creation of new products, new processes, and whole new industries.

Science and Technology Interactions

Scientific principles leading to technological breakthroughs in one field often give rise to opportunities in entirely different fields of science and technology. Technological advances, in turn, lead to further scientific advances, and thus progress continues. For example, science might generate new and improved analytic instrumentation, resulting in more sophistication in research. The advanced research induces demand and further enhancements in analytic instruments. The latter enables even greater refined research.

In the electrical industry, advances in knowledge and applications of the electrical sciences led to the creation of the lighting and appliance industry and its impact on communications. Electricity then came to be used for novel methods of medical diagnosis and treatment, leading to the discovery and establishment of the X-ray in medical practice. The X-ray, in turn, made possible the discovery of atomic physics, which ushered in the nuclear age and the release of atomic energy. World War II and the devastating consequences of the atomic bomb illustrates yet another field that has benefited from science: warfare.

Another important recent occurrence is the impact of the biological sciences on technological development. Advanced analytic instrumentation plays a key role, enabling research and discovery at microscopic and submicroscopic levels. At Stanford University, for example, researchers are using lasers, or "optical tweezers," to manipulate and examine the properties of biological molecules.[11] Now, new industries are evolving from technologically related combinations of the biological and physical sciences. Most notable is the biotechnology industry.

Advances in science have also had an impact on the location of technological innovation. Whereas in the last century the demands of technological development were simple enough to enable widespread participation by independent inventors, the increasingly sophisticated requirements of new product and process development have resulted in the growth of corporate research laboratories and "in-house" research and development. This phenomenon manifests itself in the radical shift in patent distribution over the last 100 years. In the late 1800s, less that 20 percent of patents went to corporations. Today about 75 percent do.

Licensing has come to play an important role in the development and enhancement of industrial opportunities. Nonindustrial institutions, particularly universities and government laboratories, have become significant developers of inventions. Universities now perform most of the basic research being conducted. They have essentially replaced the independent inventors as the custodians of complex science and its advancement.

Technology Transfer

Until the 20th century, much of medical practice relied on luck. Although progress had been made in clinical observation, leading to a better understanding of the symptoms of disease and of the medical properties of plants, little had happened to illuminate any underlying principles. This has been changed by modern-day advances in science and technology.

An early license that was to have a profound impact on medicine was given to an invention developed by Harry Steenbock at the University of Wisconsin. He learned how to irradiate foods and pharmaceutical products with ultraviolet light, thereby enhancing concentrations of Vitamin D. Widespread use of his invention has brought about the virtual elimination of the crippling disease known as rickets.

The patenting and licensing of Steenbock's invention was a landmark event in inventions administration. Steenbock filed his patent application in 1924. However, the University of Wisconsin was concerned about the ethics of managing patents and refused to assume the rights and administration of the invention. The prevalent concern at the time was that an institution supported by public funds should not be allowed to profit from such support. Eventually, a nonprofit organization, the Wisconsin Alumni Research Foundation, was created to administer inventions, and in 1927 it licensed Steenbock's invention to the Quaker Oats Company.[12]

A landmark license in the biotechnology area involves an invention by Stanley Cohen of Stanford University and Herbert Boyer of the University of California, which was patented by Stanford University in 1980 under an Institutional Patent Agreement with NIH, the sponsoring agency. The invention is a recombinant DNA procedure that involves a technique of introducing foreign DNA into an organism. The procedure enables the use of certain types of bacteria to produce products such as the human insulin hormone for treating diabetes. Another product of recombinant DNA is interferon, a chemical that stimulates the immune system.

The Institutional Patent Agreement granted ownership rights to the grantee institutions in return for a royalty-free license to the government. That provision has since become an element of the law governing rights in federally sponsored inventions at universities and nonprofit corporations.

About 75 companies have been licensed under the Cohen/Boyer patent. This has resulted in widespread use of recombinant DNA by industry. Each company pays $10,000 annually in minimum royalty. In 1986, royalties to Stanford and the University of California amounted to $4.6 million.

In the computer industry, an important license involved an invention by an MIT professor, Jay Forrester, also an outcome of federally sponsored research. The invention provided a magnetic core memory, a technology vastly superior to the electrostatic storage tubes of the times. It made commercial applications of computers practical. Between 1964 and 1978, licenses to industry resulted in royalty payments of $19 million to MIT.[13]

Further breakthroughs in personal computer technology were made possible by the invention of the integrated circuit, or microchip, by Jack Kilby at Texas Instruments. The microchip is smaller and less expensive than the transistor, which it replaced.[14] The development of this product continues, and again there is university involvement. Under the auspices of the New York State Science and Technology Foundation, the Center for Advanced Technology at the University at Albany (SUNY) is doing specialized research in thin-film deposition methods. Objectives of the continuing research include making PC microprocessors faster and more powerful, increasing memory densities, and further reducing the size of a microchip.

LAWS PROHIBITING ABUSE OF MONOPOLY POWER

The growth of large industrial companies in the 19th century and into the 20th century gave rise to certain practices that were perceived to be against the public interest. In general, the concern was over the ability of large corporations to create monopolistic combinations which they would then use for anti-competitive purposes. Furthermore, often the patent and antitrust laws were seen to be in conflict with each other, since the former are designed to encourage innovation by a monopoly grant and the latter condemn monopoly power as contrary to the principles of a free-market system.

It has been noted that the tendency toward corporate growth and/or concentration in high-opportunity industries such as chemicals, electricity, and recently biotechnology may be attributable to technological progress and that technological progress tends to create competitive gaps, causing successful innovators to displace unsuccessful ones. Firms in these industries watch and match each other's technological developments closely.[15]

There is evidence that firms with more than 5,000 employees are most likely to maintain heavy investment in in-house R&D, an important attribute of technological progress. But the data also reveal a limit in the effectiveness of size. Maximum scientific and engineering employment occurs in industries with four-firm concentration ratios of 50 and 55. This ratio represents the share of economic activity attributable to the four largest companies in the industry.

The data also suggest that an industry will tend to be technologically progressive if it has surmountable entry barriers and is able to induce small, high-technology entrepreneurship. Many industries have been either stimulated or revolutionized by a new entrant.

When firms in an industry become excessively powerful due to concentration and are able to control market conduct and behavior, the tendency is to attempt to maintain and exploit this power. Companies do this by engaging in anti-competitive practices and profit-seeking conduct that is contrary to the public interest. The monopolization inherent in the patent grant has the potential to violate the public interest in its inducement of such tendencies, and it is this power to subvert the public interest that antitrust laws seek to contain.

The antitrust statutes most relevant for patents are the Sherman Act of 1890 and the Clayton Act of 1914.[16] Under Section 1 of the Sherman Act, it is unlawful to combine or conspire in restraint of trade; Section 2 prohibits monopolies and conspiracies or attempts to monopolize. Section 3 of the Clayton Act forbids business practices in sales, leasing, pricing, and others that have the effect of substantially lessening competition or that tend to create a monopoly.

Notes

[1] Thornton F. Miller, *Federal Common Law in Historic U.S. Court Cases 1690-1990: An Encyclopedia.* Ed. John W. Johnson (New York: Garland Publishing, 1992), 103.

[2] Louis Orenbuch, "Trade Secret and Patent Laws," *Journal of the Patent Office Society,* Vol. 52, No. 10 (October 1970): 645.

[3] Ibid., 645-647.

[4] Bruce W. Bugbee, *Genesis of American Patent and Copyright Law* (Washington, D.C.: Public Affairs Press, 1967), 149-158.

[5] Lauren Degnan, "Does U.S. Patent Law Comply with TRIPPS Article 3 and 27 with Respect to Treatment of Inventive Activity," *Journal of the Patent and Trademark Office Society,* Vol. 78, No. 2 (February 1996): 108-120.

[6] Mitchell Wilson, *American Science and Invention: A Pictorial History* (New York: Bonanza Books, 1960), 134.

[7] Issac Asimov, *Asimov's New Guide to Science* (New York: Basic Books, Inc., Publishers, 1984), 437.

[8] D. S. L. Cardwell, *Technology, Science and History* (London: Heinemann, 1972), 153-154.

[9] *World Book Encyclopedia,* Vol. 11 (Chicago: Field Enterprise Educational Corp., 1974), 692.

[10] Mitchell Wilson, *American Science and Invention: A Pictorial History* (New York: Bonanza Books, 1974), 288-295.

[11] Howard J. Goldner, "Analytical Instrumentation," *R&D Magazine* (September 28, 1992): 58.

[12] Charles Weiner, "Universities, Professors, and Patents: A Continuing Controversy," *Technology Review* (February/March 1986): 36.

[13] Ibid., 39-42.

[14] Carl Vogel, "30 Products that Changed Our Lives," *R&D Magazine* (September 28, 1992): 42-43.

[15] F. M. Scherer, *Industrial Market Structure and Economic Performance* (Chicago: Rand McNally College Publishing Co., 1970), 376.

[16] Ibid., 422.

Chapter 3

Individuals and Organizations of Technology Transfer

Today, creation of new science and engineering applications often requires large-scale research and complex facilities. Many important technological breakthroughs that are not developed in industrial laboratories come from government, university and nonprofit laboratories. These entities have now joined the independent inventor as major sources of innovation.

The most important noncommercial parties of technology transfer, are independent inventors, the federal government, universities, and nonprofit institutions, as owners and licensors of inventions. They license their inventions to business. Technology transfer intermediaries have also emerged. Below, these parties are discussed separately, to underscore the uniqueness of each in the technology transfer process.

INDEPENDENT INVENTORS

Data on patents show that independent inventors are receiving a diminishing proportion of patents issued. However, the 20 percent of patents they do own is more than four times the combined share of universities, nonprofits, and the federal government. This portion is down from 80 percent at the turn of the century.[1]

While the argument is often advanced that inventions created by independents are typically of lower scientific content, independent inventors continue to make significant contributions to technological progress. More recently, the high-tech capability of this population is being swelled by downsizing in technology intensive industries and government sponsored laboratories.

One reason for the reduced role of independent inventors is that science-based industry, organized research, and university research programs have opened up new career opportunities for technologically talented individuals. Universities are attractive to individuals who want to engage in basic research and hope to extend the borders of scientific knowledge and understanding. Industrial laboratories offer opportunities to those who seek to use scientific advances to create tangible products. In the past, these opportunities either did not exist or were too few to absorb interested people as employees. The many creative individuals who now work for research establishments would most probably have been among the ranks of independent inventors in an earlier era.

Another incentive for inventors to become employees rather than independents is that many universities and research establishments allow inventors to acquire ownership rights to their inventions. Recently, the federal government has also become more flexible with respect to granting rights to its employees who create inventions.

Until the emergence of the modern corporation as a separate entity in the latter part of the 19th century, business sprang largely from the creativity of the individual. Notable examples are Thomas Edison and Thomas Howe, both of whom were recipients of royalties under patent rights and technology transfer. Examples were also given in Chapter 2 of the emerging large corporations buying up patent rights to consolidate their market power. Most of those rights were purchased from independent inventors.

Unfortunately, reliable data are not available to show the extent to which businesses are created by independent inventors who retain their ownership of the patents or by the sale or licensing of patent rights to an existing corporation or venture capitalist who then creates a new company to market the new technology. However, the costs of obtaining and maintaining a patent are such that few would undertake the effort unless they believed that the invention could be a commercial success.

THE FEDERAL GOVERNMENT

The federal government both supports and performs research and development (R&D) activities that create new technology. Through policies and legislation governing technology transfer, the government helps provide incentives for new inventions and insures that the public benefits from inventions that the public has sponsored.

The development of the government's widespread participation in R&D is largely the consequence of two world wars and the subsequent tensions of the Cold War. These experiences amply demonstrated the importance of science and advanced technology for national defense and the security of the nation. The Great Depression was also instrumental in changing public attitudes about the government's role in promoting scientific advances.

While science and technology policy had its roots in the Constitution, until the 20th century, government participation in science was limited. However, during the earlier years it did take some actions that were to later have major consequences for technological development. It created the Marine Hospital Service in 1798. In 1912 the Marine Hospital Service changed its name to the Public Health Service (PHS).[2] The PHS is the present day parent of the National Institutes of Health. Later, in the Morrill Act of 1862, the government established Land Grant Colleges, which provided people in states across the nation with instruction in agricultural sciences and mechanical arts. Military science was later added to the curriculum. Subsequent legislation ensured a continuing flow of support to these Land Grant colleges and universities. Today, some of them are renowned centers of research.

The threat to national security posed by World War I (WWI) initiated the first recourse by government to the nation's research resources for a major public purpose. The war effort drew on the research capabilities of business, foundations, and universities. To coordinate their work, the government instituted the Council on National Defense, the Naval Consulting Board, the National Academy of Sciences, and the National Research Council. This strong federal support for science expired after the war and did not reassert itself until World War II (WWII).

In the years between the wars, government action for the public good became pronounced under President Franklin D. Roosevelt, who endeavored to use the government's power to lift the country out of the Great Depression. The British economist John Maynard Keynes greatly influenced new thinking about government's ability to influence economic affairs. That new thinking, coupled with the desperation of the times, helped shift public opinion towards a greater acceptance of the role of government in people's lives. The more activist role culminated in historic social security legislation and the Full Employment Act of 1946.

Mobilization for World War II occurred within this climate of increasing public acceptance of broader government action. Again, business, government, and universities were brought together in the nation's defense. The Office of Scientific Research and Development was created in 1941. Atomic energy for military and civilian purposes was supported under the Atomic Energy Commission. Basic research to advance military technology was supported by the Office of Naval Research.

After the war, during the 1950s, the National Aeronautics and Space Administration (NASA) was born in response to the launching of *Sputnik* by the Union of Soviet Socialist Republics (USSR). Advocates also came forth, looking to science for ways to improve the economy, the health of the people, and the general welfare. By the end of the 1950s, unprecedented rates of growth were evident in those federal agencies that sponsored scientific research.

The National Science Foundation (NSF) was created in 1950 to support basic research. The National Institutes of Health (NIH) were established in 1948 to conduct and sponsor medical research. They were an agency of the Public Health Service, which in turn had become a division of the Department of Health, Education and Welfare. Today, many arms of the federal government promote research and development enterprises, including the departments of State, Defense, Agriculture, Commerce, the Interior, and Transportation; the Public Health Service; NASA; and the National Science Foundation.[3]

Until recently, inventions arising from government research were simply released to the public, free for all to practice. Proponents of this policy argued that since these inventions were the result of public support, they should belong to the public. Exclusive benefit should not accrue to any one individual or corporation at public expense. Thus, the use of new inventions should be on a nonexclusive basis. Little commercial development of new inventions by business came of this policy, however, due to the high risk of introducing a new product and the fear of an unproductive investment.

As the industrial countries recovered from World War II, the United States was increasingly challenged in world markets and began to lose its competitive advantage. The result was a shift in government policy from one of government as prime customer of federally sponsored technology to one of partnership with the private sector.[4] The old assumption that somehow government patents would automatically find their way into commercial use was abandoned in favor of a policy that actively sought to encourage such use. This change is evident in significant pieces of legislation enacted since 1980.

The cornerstone acts for technology transfer from the nonbusiness to the business sector are the University and Small Business Patent Procedures Act of 1980 (commonly known as the Bayh-Dole Act) and the Stevenson-Wydler Technology Innovation Act, enacted in 1980. The Bayh-Dole Act enables researchers under federal contracts to acquire rights to the inventions they create. They may then grant exclusive licenses for these inventions, thereby making such technologies more attractive to industry. The Stevenson-Wydler Act seeks to encourage technological transfer from government laboratories to industry, under licenses, employee spin-offs, and limited partnerships. This activity has been further strengthened by the Consortium for Technology Transfer Act.

In 1992, the federal government provided roughly 28 percent of the support for research and development in industry and an estimated 57 percent of the research funds expended by colleges and universities. An estimated 11 percent of the nation's R&D was performed by federal employees in federal agencies.[5] The percentage was about the same for basic and applied research. This massive scale of government involvement in the nation's R&D endeavor is spread across virtually all the federal agencies. Under the Small Business Innovation Development Act, the Small Business Innovation Research Program (SBIR) requires agencies with research budgets exceeding $100 million to set a certain percentage of their research aside for small business.

The government also operates Federally Funded Research and Development Centers (FFRDC), some of which are administered by industry at certain universities. Those administered by universities performed 3.2 percent of the nation's R&D in 1992. However, their share of basic research was 10.3 percent, bringing the combined total for universities and colleges and FFRDC's to 59.4 percent of the nation's basic research effort that year.[6]

UNIVERSITIES AND COLLEGES

Instruction, research, and the dissemination of knowledge through publications and public service are the basic functions of colleges and universities. Every nation depends on the education and skill level of its population, and/or its scientific output to induce and sustain technological progress. Each nation looks to its institutions of higher learning as prime agents for advancing the education and skill levels of its people.

Instruction

The understanding, adoption, application, and continuing advance of science and engineering principles are critically important for modern technological development and progress. Most notably,

> The rapid spread of modern education must have been a basic element in increasing the capacity of developed nations to exploit and contribute to the available stock of tested and useful knowledge. It provided a common language for increasingly large groups . . . and thus a widening basis for sharing in and contributing to a common body of knowledge and technique.[7]

Colleges and universities have steadily raised educational levels in this country, continuing a trend begun with the founding of the nation. In 1900, the number of bachelor or first professional degrees conferred per 1,000 persons 23 years old was 19. It was 81 in 1940, and 223 by 1970.[8] In 1992, more than 22 percent of persons 25 years old and older had completed four years of college or more. In 1960, this ratio was less than 8 percent.[9] As another measure of continuing growth, in 1989 the number of scientists and engineers employed totaled 949,300, as compared with 543,800 in 1970, an increase of 75 percent.[10]

Research

The importance of research has increased greatly since the early part of the century. Research, with its manifestation in publication has become a condition of continuing employment at most colleges and universities today, and the popular saying, "publish or perish" is a reality for most academics. There is also growing documentation of the close relationship between research, publication, and technological development.

Early important contributing factors toward a public perception of the importance of universities in research were collaborations with industry, endowments from foundations, and donations from wealthy individuals. Support from the federal government was small and directed towards a few applied fields. However, by inviting universities to participate in a partnership with government and industry, for the purpose of harnessing technology for World War I, the federal government affirmed the role of universities as prime performers of basic research. Similar action is noted above with respect to pursuit of war objectives, during WWII, under the Office of Scientific Research.

Colleges and universities now perform about half of the sponsored basic research of the nation (49.1%), more than the combined performance of industry (21.3%) and the federal agencies (11.1%).[11] This performance is reflected in contributions to the research literature, in which more than two-thirds of publications in the most influential science and technology journals are articles authored by academics.[12] The significance of these statistics for technological development is revealed in data that show a narrowing of the gap between science and technology.

In one study statistics are presented demonstrating the dependence on science of technological development in industries. The authors use U.S.-granted patents as a measure of technology and the "other references" appearing on the front page of the patent as a measure of the contribution of science. They use citations of scientific works per patent and the time lag between the date of publication and its citation as the important characteristics to be quantified.

In biotechnology, both patents and other publications show a peak of two to four years from the date of publication of the cited article to the date of citation. Also evident is a declining median age for citations in all patents studied. The greatest dependence on the scientific literature appeared in patents for drugs and medicine, where 72 percent of the citations were scientific research journals, followed by scientific instruments, computing and communications.

Patents in chemistry and allied products, which contain drugs and medicine, relied the most on basic as opposed to applied research journals; other fields generally cited applied science journals. Advances in transportation and machinery ranked lowest in their dependence on scientific journals. They referred more often to nonscientific publications, such as technical reports, specifications, and disclosures.[13] Such findings are indicative of a blurring of the distinction between science and technology and of the importance of the contributions of universities and industry to technological progress.

Public Service

University do patenting and licensing under the public service category. The advance of these institutions and their acceptance into the arena of patenting and licensing was tentative in the pre-WWII era. It grew significantly after WWII, however, and has become particularly pronounced since passage of the Bayh-Dole Act in 1980. Today, the patenting and licensing of inventions developed by university faculty is a common practice. The scale of this activity is evident in a 1993 survey of 250 members of the Association of University Technology Managers. Some 85 percent of the respondents reported that their universities were pursuing patents. In 1934, only 18 universities did so.[14] The university respondents also reported receiving 1,307 U.S. patents in FY1993, with licenses and option agreements executed numbering 1,737.[15]

Universities and colleges also become agents in technology initiatives of state governments. Typically, a state's interest in technology is to promote its own economic development and the growth of high-technology industries within its borders. Accordingly, it sponsors programs to promote close ties between universities and industry. An example is the New York State Science and Technology Foundation, which was created by the State of New York as a public corporation for this purpose. Among its programs, it operates 13 field-specific Centers for Advanced Technology (CAT) at the top universities across the state. Their mission is to facilitate technology transfer from the host institutions to private industry.

Another important public service with technology transfer implications is the consulting to industry done by faculty. Universities commonly allow faculty consulting time of up to one day a week. The practice enhances the science and engineering content of industrial production as well as awareness of industrial research and development needs among faculty. The creation of communication networks through consulting relationships have the potential to improve industrial research funding levels to the university.

Other tangible benefits to the colleges and universities accruing as a result of relationships with the industrial sector include donations of equipment and materials, and the placement of their graduates in career paths.[16] Similar benefits accrue from active licensing programs that promote communications between university faculty and industrial technology representatives.

University/Industry relationships through the years, however, have not been without controversy. Many of the concerns endure to this day. One source of controversy springs from the divergent objectives of the parties. Industry research is driven by the need for specific results developed under conditions of secrecy. University research is based on disinterested inquiry, accompanied by open communications and public dissemination of research results. The unique role of colleges and universities in education and the advancement of pure science for the public benefit relies on the free flow and exchange of information.

The influence of interest in economic gain from the results of research is also a reason for concern. Critics argue that interest in gain diverts the attention of faculty from instruction and encourages an environment of secrecy. They fear that the several allegiances of university scientists, as recipients of industrial research support, as consultants to industry, as stockholders or equity participants, compromises their responsibilities to the public as expert witnesses in increasingly technical policy issues. Critics further argue that the goal of universities as centers of instruction may be compromised by their immediate interest in financial gain from research and that the research itself may be diverted to short-term commercial interests.

THE INDUSTRIAL SECTOR

Businesses provide the vehicle by which the eventual outputs of research, inventions, become translated into technological progress through commercialization. Industrial research and development is an important aspect of this process. By the turn of the century, the importance of science for innovation was increasingly being recognized by industry. The research laboratory concept pioneered by Edison spread as other companies began to conduct research. By the early 1900s, General Chemical, Dow, DuPont, Goodyear, Eastman Kodak, and American Cyanamid had established centralized research laboratories. Today, survival in high-tech industries is not possible without a commitment to scientific applications.

Personnel and Infrastructure

Employees trained in the methods of science and engineering today ensure continuing advance in the technological sophistication of products and processes. Industry sought people educated in these fields in the early 1900's. Scientists with Ph.D.'s were in high demand. Besides recruiting them from universities, companies paid universities to assign special industrial projects to their students.

Companies also look to universities for their needs. They sponsor graduate students and contract research at universities across the nation. More recently they have also begun to make multimillion-dollar awards to university laboratories, in support of a particular area of basic science. One example is a 12-year agreement executed between Harvard and Monsanto to support specialized research in biochemistry. Another example is a 10-year agreement between Exxon and the Massachusetts Institute of Technology (MIT) involving research in combustion science.

Besides a commitment to research and development, technological progress needs an infrastructure. The adoption of scientific and technological advances greatly influences the development of a commercial infrastructure, and there exists a codependence between business, research and infrastructure. By the turn of the century, the companies of the electrical industry had already made significant contributions to the development of the nationwide infrastructure of communications, transportation and power generation. Growth was particularly stimulated by the emergence of mass markets.

Science, Patents and Business Competitiveness

Companies contributed to and took full advantage of the new opportunities, applying science and engineering advances and using the protections of the patent system. This is particularly evident in the business strategies of the companies that came to dominate the electrical industry.

To remain competitive, corporations used the rights conferred under the patent system to achieve a strong market position. This was not unusual as a business practice. In order to gain control of new technologies, a company would develop and/or acquire patents to as many significant inventions in its area as it could. General Electric (GE), Westinghouse, and American Telephone & Telegraph (AT&T) established R&D and patent departments, to that end. They produced patents internally, purchased them from the outside, and acquired them by buying up competitors and by merger, strategies that continue to this day as standard business practices in the business world.

The chemical industry provides another striking historical example of the importance of patents in the development of United States industry. While the chemical industry had undergone significant development by the close of the 19th century, it did not have companies with commanding market positions comparable to the GE's or ATT's of the electrical industry. Such industry giants did not emerge until after World War I. This was due partly to the control of major patents by German companies. According to one study, German companies owned 98 percent of the applications in 1912.

After the war, however, the German-owned patents were seized by the US government and released to the highest bidders. Small companies protested the inequity, and a Chemical Foundation was created, charged with licensing the German patents on a nonexclusive basis. But the beneficiaries continued to be the strongest chemical companies, including DuPont, General Chemical, Bausch & Lomb, and the Newport Company.[17]

Few industries today lack a science base. As stated by James Conants,

> . . . science emerges from other progressive activities of man to the extent that new concepts arise from experiments and observations and new concepts in turn lead to further experiments and observations. . . . The texture of modern science is the interweaving of fruitful concepts.[18]

With the maturing of the science and technology relationship, the term "high-tech" has been coined to differentiate industries according to their respective dependence on research and development and scientific progress. Relevant examples of industries engendered by this relationship are the chemical and electrical industries cited above and more recently, pharmaceuticals; computers; electronics; medical instruments, equipment, and devices; and biotechnology.

High-tech industries may also be characterized as high-opportunity industries: industries that manifest high growth and/or concentration. Such growth and/or concentration is said to be attributable to technological progress, which creates competitive gaps that enable successful innovators to displace unsuccessful ones.[19]

The relative performance of firms in an industry is manifest in patent statistics, which correlate with a number of important indicators. Patent count and the frequency of citation of patents provide revealing insights into the relationship between patents and corporate strength.[20] A favorable correlation occurs between patent count and various aspects of R&D performance, including new drug registration and approvals, output of research papers, expert opinions, and drug composite output score.

Financial performance correlates highly with citations received per patent and with technology concentration of patents. A patent on a major technological breakthrough would lead the patenting company to patent succeeding inventions relating to and citing that patent, consolidating its patent position as it seeks to ensure a maximum return on the technology with minimal interference from competitors. In the meantime, competitors and imitators are drawn to the opportunities of the new industry. The influence of the original patent is revealed in its citations on subsequent patents.

These findings demonstrate the importance of patents in industry, their relevance as profit generators, and in particular, the potential significance of third-party patenting as a business strategy for technological progress. A previous section noted the relevance of third-party research and publication for patenting. The technologically progressive business supplements this science source and profit opportunity with licensing activities. Third-party owners of inventions and patents are often turned to as licensors for technological breakthroughs, new products or processes, and/or the maintenance of competitive advantage.

NONPROFIT RESEARCH INSTITUTIONS

Nonprofit institutions are a mixed population. They have traditionally been, and continue to be a powerful player in the development and applications of science. Their activities include the conduct of research and collaborative work with universities and hospitals, where they may also fund research projects. They facilitate interactions among companies as trade associations, as well as between the academic and industrial communities.

Philanthropy is a principle determinant in the creation of many of them. Some nonprofits spring from industrially motivated arrangements with universities. Others initially created as separate entities combine with universities or become departments of universities. Many remain as independent entities. Today, nonprofits number in the hundreds and vary in size from very small to the very large.

Besides emerging as respected institutions of research at the turn of the century, nonprofits have been instrumental in defining the organization of collaborative industry/university relationships. They pioneered in areas of contract research and personnel training for industry. Services for a fee became a major undertaking under the Mellon Institute, which sponsored the creation of a department for industrial research at the University of Pittsburgh, initiating the fellowship program for new Ph.D's.

The Mellons are an example of business people with immense fortunes who went on to create institutions for the pursuit of industrially relevant science and basic research. This institute later merged with the Carnegie Institute of Technology to form the Carnegie-Mellon University. Philanthropists established the Arthur D. Little Foundation, the Battelle Memorial Institute, and the Armour Research Foundation, all of which became independent research institutes with a focus on contract work.[21]

A nonprofit may be specific to a given industry, technology or public need. Examples of the first include the Institute of Paper Chemistry and the Institute of Rubber Research. However, such industry-specific interest can have technological implications for multiple industries as evidenced by the Semiconductor Research Cooperative. Its support of advances in microelectronics impacts computer manufacturers, instrument makers and the telecommunications industry.

Public purpose nonprofits may have technology-specific orientations, such as health and medical sciences at the Massachusetts General Hospital, Brigham and Women's Hospital and the Scripps Clinic & Research Foundation. Others specialize in defense contract research. Among these are, the Jet Propulsion Laboratory, the Lawrence Radiation Laboratory, and the Institute for Atomic Research.

Available statistics show that most of the research conducted by the nonprofits is located at the independent research institutes. The nonprofit federally funded research and development centers (FFDC's) rank second. A significant amounts of research is also conducted at the voluntary hospitals. By far the greatest sponsor of the research at the nonprofits is the federal government, although industry supports most of the sponsored work undertaken by trade associations.[22]

TECHNOLOGY TRANSFER INTERMEDIARIES

An industry for helpers of inventors with their marketing, licensing, and patenting needs has emerged. The participants exist as either individual consultants, for-profit or nonprofit entities, and services range from help with the basics, to highly specialized, to complete transfer technology management. A search on the Internet, using Infoseek as the search engine and the combination keywords "invention marketing" reveals a range of help for the invention owner, including warnings against scam artists.

Patent attorneys are available as an absolute necessary help in the highly specialized category of patenting. Any of the core services in technology transfer management, marketing, patenting, and licensing, can be contracted out, if one is willing to pay for them. Payment forms vary, from fixed amounts for defined services to percentages of income from licensing to combinations of the two.

Entities have also arisen to assist in the technology transfer of institutionally-produced inventions. The services they provide may be institution-specific, or they may be available broadly to any university or college seeking help with marketing and patenting. These corporations generally do not cater to the individual inventor having no institutional affiliation.

The technology transfer offices housed within an institution providing multiple services to a university is one type of arrangement. The Research Foundation of State University of New York (RF/SUNY) is an example. Among the charges of RF/SUNY are the marketing, patenting, and licensing of SUNY faculty inventions that are managed by its Technology Transfer Office. Inventions are administered under the Inventions and Patent Policy of the University. Faculty are required to assign their rights in inventions to the University as a condition of employment.

If the technology is the result of externally sponsored research, title of the invention goes to RF/SUNY. The Patent Policy further requires that faculty inventors receive 40 percent of gross royalties in return for the assignment of their rights. However, inventors may reacquire their rights if the RF/SUNY decides not to patent or market the technology, subject to sponsor regulations.

Another form of nonprofit entity is the Wisconsin Alumni Research Foundation (WARF), founded specifically for that university for the purpose of not grant, but invention administration. Faculty of the University of Wisconsin have the option of retaining rights and managing their own inventions. An option available to them is the service of WARF. If they choose to use these services, they must then assign their ownership rights to WARF.

Research Corporation, the parent corporation of Research Corporation Technologies (RCT), was founded as a nonprofit organization to acquire and administer university-produced inventions. The present-day operating entity is the RCT, a tax-paying entity that continues the original mission as a spin-off company. RCT deals primarily at the institutional level, contracting directly with interested universities and colleges. The terms involve assignment of ownership rights from the university to RCT. RCT then retains a percentage of gross income from licensing, and assumes the costs for marketing and patenting.

Examples of for-profit entities providing full services are, British Technology Group USA and Competitive Technologies, Inc. BTG is similar to RCT in requiring full assignment of ownership rights in inventions, and in assuming costs for patenting. As compensation they keep a portion of royalties from licensing. Competitive Technologies does not require that it be given title to the invention, however it also does not assume patenting costs. Others are, Arther D. Little Enterprises, Inc., and The Western Patent Group.[23] These entities also feature an internet homepage describing service offerings.

Technology transfer agents providing indirect services are also available. These entities leave the direct communications about inventions available for licensing to the parties. Instead, they serve to make the opportunity known. Owners of databases or publishers of newsletters listing such technologies are examples. Firms that organize technology transfer conferences are another example of entities providing indirect services. Here, available technologies are featured and the parties of the technology transfer transaction meet directly to determine if there is mutual interest. The various services are available either on a subscription basis or for a fee.

Notes

1 Jacob Schmookler, *Invention and Economic Growth* (Cambridge, Massachusetts: Harvard University Press, 1966), 26.

2 *World Book Encyclopedia*, Vol. 15 (Chicago: Field Enterprise Educational Corp., 1974), 757.

3 Krishna Mathur and Ralph Sanders, "Science and the Federal Government," in *Science and Technology: Vital National Resources.* Ed. Ralph Sanders. (Maryland: Lomond Systems, Inc., 1975), 87.

4 Richard Brody, *Effective Partnering: A Report to Congress on Federal Technology Partnerships* (U.S. Department of Commerce, Office of Technology Policy, April 1996), 23-35.

5 National Science Foundation. *National Patterns of R&D Resources* (1992), 3-5.

6 Ibid., 18.

7 Simon Kuznets, *Modern Economic Growth: Rate, Structure, and Spread* (New Haven, Connecticut: Yale University Press, 1972), 289.

[8] 93rd Congress, 1st Session, House Document. *Historical Statistics of the United States, Colonial Times to 1970* (U.S. Department of Commerce, Bureau of the Census), 385-386.

[9] U.S. Department of Commerce. *Statistical Abstracts of the United States 1993,* 153.

[10] Ibid., 602.

[11] National Science Foundation, *National Patterns,* 18.

[12] National Science Board, *Science Indicators. The 1985 Report,* 92.

[13] Francis Narin and Dominic Olivastro, "Status Report: Linkage Between Technology and Science," in *Research Policy,* Vol. 21 (1992), 237-249.

[14] Charles Weiner, "Universities, Professors, and Patents: A Continuing Controversy," *Technology Review* (February/March 1986), 39.

[15] Association of University Technology Managers, Inc. Conducted by Diane C. Hoffman, Inc. *The AUTM Licensing Survey, Executive Summary and Selected Data, Fiscal Years 1993, 1992, and 1991* (Connecticut: Association of University Technology Managers, Inc., 1994), 11.

[16] National Science Board, *Science Indicators,* 107.

[17] David E. Noble, *America by Design—Science, Technology and the Rise of Capitalism* (New York: Alfred A. Knopf, 1977), 10-16.

[18] Taken from Morris C. Leikind and Wyndham Miles, *The Nature of Science and Technology, in Science and Technology: Vital Resources.* Ed. Ralph Sanders. (Maryland: Lomond Systems, Inc., 1975), 2.

[19] F. M. Scherer, *Industrial Market Structure and Economic Performance* (Chicago: Rand McNally College Publishing Co., 1970), 378.

[20] Francis Narin, Elliot Noma, and Ross Perry, "Patents as Indicators of Corporate Strength," *Research Policy* 16 (1987): 143-155.

[21] Arnold Thackray, "University-Industry Connections and Chemical Research: An Historical Perspective" in *National Science Foundation: Selected Studies,* 213-233.

[22] National Science Foundation, R&D Activities of Independent Nonprofit Institutions, 1973, *Surveys of Science Resource Series,* NSF 75-308, Table 8-9.

[23] Mention here of the technology transfer entities is purely informational and does not represent an endorsement by the author. Before selecting an agent the technology owner is advised to investigate the reputation of the entity, or individual offering the services.

Chapter 4

Elements of Patent, Trade Secret, and Antitrust Law

The ability of the parties of technology transfer to enter into licensing arrangements is dependent on the enforceability of intellectual property rights, and the privilege accorded commercial practice under such rights. Under such rights the licensor, who holds the property rights is able to grant permission to the licensee to use the rights for commercial purposes. Intellectual property rights are enforceable under the trade secret and patent laws.

Secrecy governs invention rights under trade secret law, whereas under patent law rights to the owner are tied to public disclosure. Unlike a trade secret, the rights in a patent are conveyed under a physical document, and they have a defined life. To secure Letters Patent, as the document is called, certain criteria and subject matter must be satisfied. Because trade secrets and patents convey rights that tend to exclude others from competition, they sometimes come into conflict with U.S. anti-monopoly or antitrust law.

THE INVENTION AS A TRADE SECRET

Trade secret law does not require the owner of an invention to seek a patent in order to acquire intellectual property rights, nor is the owner required to disclose the new technology to the public. The owner can legally exploit know-how commercially while keeping it secret. Protection under trade secret law depends on secrecy, which in turn depends on the care the owner takes to protect the invention from unwanted disclosure.

Resource to trade secret law is lost if the secret becomes known, whether or not the disclosure was made by the owner of the secret. The owner who wishes to disclose an invention is free to make it known to everyone. However, unless the invention is protected by a patent, the law can do little to prevent its subsequent use by others. Owners who desire the protection of legal recourse must exercise caution with regard to disclosure of their inventions.

Trade secret rights are generally not available if the new technology is readily deducible by others or is obvious. The trade secret holder has no legal remedy against independent development or "reverse-engineering," whether such development is done in secret or with knowledge of the trade secret holder. Even if the trade secret holder has the right of legal recourse against an offender who makes the information available to others, the trade secret holder has no legal recourse against these others using the information, provided these others are innocent third parties.

The protection of the law is available to the owner who is willing to disclose the new technology as well as to the owner who wishes to keep the invention a secret. In the case of disclosure, the purpose would normally be to locate a buyer or a licensee, and the law protects the owner from unfair appropriation and use of the invention by recipients of the information. When the owner chooses to keep the invention from others, as a trade secret, the intention would be exploitation by the owner. In this case the invention usually confers an advantage in trade that the owner can only enjoy if the invention remains unknown to competitors.

When the holder of a trade secret discloses it to another, the question under trade secret law is, does the person learning the secret have a duty to refrain from using or disclosing it to others? Important factors in answering this question are the manner in which the learning took place and the potential injury disclosure would cause to the holder of the secret.

The holder who wishes to maintain the secrecy of the invention has recourse against a third party who discloses or uses the trade secret information if improper means were employed in learning the secret information, or if this information was obtained in confidence from the owner. A relationship of trust, as that of an employer and an employee may result in an acquisition of trade secret information. In this instance, an employee or former employee is liable if he or she discloses the information or uses it to further his or her business interest.

A protectable disclosure is one made under an express contract. The disclosure occurs in confidence. Such a contract would further stipulate that nonpublic information must be held in confidence for a specified period of time.

Putting a company on notice that an invention disclosure is about to be made and giving it an opportunity to reject the disclosure creates the conditions of an implied contract, particularly if the recipient ignores the notice and subsequently uses the disclosed invention. However, the disclosed information must not already be in the public domain.

The benefit of trade secret is in the advantage over competitors that the secret information confers on the holder. This advantage is jeopardized, or lost if the information becomes known. This loss constitutes the injury and unfairness to the holder. The injury and unfairness are brought about by disclosure and subsequent use of the trade secret, and resulting benefits to competitors at the holders expense, in the loss of a trade advantage. By protecting the holder, the opportunity for a competitive advantage, a condition for success in free enterprise is preserved.

PATENTS

The patent is a teaching document. In exchange for instruction of the public in the technology, the owner is given a legal right to exclude others from practicing the invention for commercial purposes. Under this law, the owner can set the conditions under which others can practice the technology. With the exclusive right to practice the invention, or license others for financial gain to the owner, inventorship is rewarded and thereby encouraged.

Patent Term

Until recently, the life of a U.S. patent was 17 years from the date of issue of the patent. This term changed on June 8, 1995, as a result of legislation implementing the General Agreement on Tariffs and Trade (GATT). The term is now 20 years from the filing date of the patent application, irrespective of the number of years it takes to process the application.

However, if the application was still pending on June 8, 1995, the resulting patent will have a life of 17 years from the date of issue or 20 years from the date of the application, whichever is greater. A patent obtained from a succession of related applications is given for 20 years from the date of the first, or priority, filing.

The new law allows for extensions of the patent term up to an additional 5 years to compensate for delays caused by interference proceedings, secrecy orders, and appellate review. An extension of up to 5 years may also be obtained under the Drug Price Competition and Patent Term Restoration Act of 1984. This is to compensate the holder for market entry delays due to regulatory review, and applies to both product and method claims.

The new law also provides for a provisional application. This provisional application is to be distinguished from the "regular" or "complete" application noted above. The provisional application does not begin the life of the patent. However, it can add a year of protection. To enjoy the benefits of the provisional application, an applicant must file a complete application within 12 months after the date of the provisional filing.

Conditions for Grant of a U.S. Patent

Patentable material is defined under Sections 101-103 of the U.S. Patent Law (Title 35, United States Code). The subject matter must be novel, must have utility, and it must be nonobvious. For subject matter, the invention must be a process, a machine, a composition of matter, manufacture, or any improvement made to them.

Novelty: The condition of novelty means that the patent is issued to the applicant who first invents the subject matter. The subject matter must not have previously been invented by someone else or known by others in the United States. It must not have been described in an issued patent or a publication already accessible to the public. The invention must not already be in a patent application in the United States or in foreign countries.

If the invention becomes known to the public, the patent must be applied for no later than one year after such disclosure. This provision is known as the "grace period" for filing. Also, if an application has been made in a foreign country, the U.S. filing date should be within 1 year of that filing in order to claim priority. In the event of a dispute, the inventor may be required to demonstrate the invention date by producing dated laboratory notebooks containing his or her witnessed signature.

Most foreign countries will not issue a patent if the application occurs after a public disclosure. However, the Paris Convention provides that if there has been a prior filing in the United States, the U.S. filing date becomes the "priority date," provided the foreign application is filed within 1 year of the U.S. filing date and priority is claimed. The Paris Convention is one of the international agreements regarding intellectual property in which the United States participates.

Utility: To be eligible for patent coverage, the subject matter of an invention must be a process, machine, composition of matter, or manufacture or an improvement thereof, that is useful. The invention must be operable and must serve some useful purpose. For new drugs, a minimum level of safety and efficacy might be required. The purpose for which the invention is useful must be described in the patent, and a person skilled in that art should be able to reproduce at least one of the results claimed. However, production of a marketable commodity is not required, although market acceptance can be used to establish utility. If the asserted utility is speculative, the patent must contain a basis for the supporting belief.

Nonobviousness: Although the invention may satisfy the conditions of novelty and utility, it may be considered nonpatentable if its departure from the existing art is not enough to make it nonobvious to one skilled in the art viewing the entire literature.

This determination must be supported by facts concerning the prior art, the difference of the invention from this art, and the pertinent level of ordinary skill. The Examiner in the Patent Office, whose job it is to determine the issues of obviousness and nonobviousness from the facts concerning the prior art and the invention, may have more or less skill than one of ordinary skill but applies the test using a hypothetical person of ordinary skill in the art.

Statutory Subject Matter

Process: This term may be taken to mean art, method, or mode of operation and includes a new use of a known process, machine, manufacture, or composition of matter. The process must be executable on a machine, manufacture, or composition of matter in one or more steps and must produce a physical result.

A patentable process could involve the application of old steps to new subject matter, new sequences of performance for old steps, combinations of old steps with new steps, or entirely new steps. The steps of the process and the material acted upon define the patentable subject matter. For example, a new step might be a new starting material in a chemical process.

Machine: The term *machine* applies to both hand-operated and automatic mechanisms. It also means devices, engines, or apparatus that results in a observable output when activated. As with the steps of a process, the machine may be unique by virtue of a part or parts and/or how the parts are arranged to produce the intended result. A patentable machine may be either the total entity or its component or components.

Composition of Matter: While the term *machine* alludes to combinations of parts or components producing a desired end, the term *composition of matter* pertains to ingredients comprising chemical or physical elements or properties combined as compounds or mixtures that produce a defined effect. Patentability is defined by the uniqueness of the ingredients, by the novelty of their combination, or by both.

Manufacture: *Manufacture* is a term generally used to embrace inventions that do not fall under the other statutory categories. It includes products as varied as building structures, sound recordings, and organisms resulting from or substantially changed by human genetic engineering.[1]

Computer Software

Software relates to computer source code, object code, procedures and documentation associated with the operation of a computer, its performance, or the production of an output by the computer.

However, there are categories of software that are generally not patentable. A patent will not be granted for a mathematical algorithm or for the manipulation of an idea that has no practical value. Also, data structures, information, or computer programs that do not implement processes on the computer or in any way contribute to the operation of the computer are generally not considered patentable. Abstract ideas, laws of nature, and natural phenomena such as electricity are also not patentable.

Utility (usefulness or functionality) plays a crucial role in determining patentability for a computer software. Patentable and nonpatentable subject matter is separated by a test of functionality. This test may be satisfied if a data structure is used in a memory circuit to perform a practical application or if a computer program is used to implement a computer process. Possible statutory subject matter for the patents in these two instances would be manufacture and process, respectively.

However, information incorporated in an item of hardware does not automatically give rise to patentability. For example, in the case of music, art, or literature that is communicated by computer, the value resides in appreciation by a person, not in an internal contribution of the information to the operation of the computer. Also, use of such information by the computer does not result in a physical transformation.

The acceptance of computer software as statutory subject matter for patenting is relatively recent. In 1964, software was determined to be unpatentable by the United States Patent and Trademark Office (PTO), because it was considered to be "creations in the area of thought." In 1968, the PTO determined that software combined with an apparatus was to be considered patentable. However, this decision was later rescinded, to be reinstated only in 1989, when the PTO announced that it would accept physical processes implemented by computer codes and algorithms.[2]

Developers of software have not been without legal rights in the absence of patent protection however. Software was and continues to be protectable by copyright law. Copyright protects the expression of an idea, but not the idea itself, as a functional concept. It prohibits third parties from making copies or reproductions of the expression of the idea. Unlike a patent, copyright provides little protection against someone who might have developed the idea independently or who might have arrived at the same functions expressed in the idea by "reverse engineering."

Structure of a Patent

A patent, (or Letters Patent) is divided into several sections, which make up the general format required of patent applications under the Patent Act of 1952. The descriptive material in the patent is arranged under the headings: (a) Title, (b) Abstract, (c) Background of the Invention, (d) Summary of the Invention, (e) Description of Specific Embodiments, and (f) What Is Claimed Is.

Besides the Title and Abstract, the first page must contain the patent date and number and a drawing that might be relevant. The first page also carries the names of the inventor; patent attorney; Examiner; assignee, if any; related patent applications (indicated if abandoned for others); priority dates; serial numbers; and the prior art cited by the Examiner. The patent contains bracketed numbers throughout. These are elements of an internationally accepted code by which segments of the patents can be identified, regardless of language.

Title: While the title introduces the technology, it is generally not revealing. The Patent Office tends to be flexible in accepting a proposed title, provided the title does not entirely misrepresent the invention.

Abstract: The abstract is a brief synopsis of the invention. While it is not a part of the disclosure, it serves as technical statement of the technology.

Background of the Invention: This section describes the prior art relating to the invention and states the problem to be solved.

Summary of the Invention: The summary expands upon the abstract and provides a brief description of the object and benefits of the invention without elaborating on its technicalities.

Description of Specific Embodiments: This section describes the various features of the invention in technical terms, including its usefulness or utility, and provides working examples. The specifications must be sufficiently thorough as to enable a person skilled in the art to reproduce the invention and practice it. To that end, the patent also specifies the best mode for carrying out the invention. In so doing, the specifications represent an interpretation of the patent's claims.

In the field of microbiology, the "best mode of carrying out the invention" usually includes the availability of a culture. A patent applicant is thus required to deposit an appropriate culture in a patent depository. The Northern Regional Research Laboratories of the Department of Agriculture (NRRL) in Illinois accepts cultures of nonpathogenic bacteria and fungi only. The American Type Culture Collection (ATCC) in Maryland accepts all cultures of bacteria, fungi, algae, protozoa, cell lines, hybridomas, oncogenes, plasmids, viruses, plant tissue cells, seeds, and animal embryos (broadly defined as microorganisms).[3]

What Is Claimed Is: The claims are the legal description of the owner's property rights. An important reference for interpreting the claims is the "Description of Specific Embodiments." All that is claimed is described in the specification. However, the specification is not the only basis for interpreting the claims.

Claims may be broad or narrow in scope. Broad claims afford protection in that a third party seeking to use a similar technology will generally have difficulty doing so outside the scope of the claimed invention. It is much easier for a third party to avoid infringing on a narrow claim. In practice, the claims usually include one or more independent claims, and dependent claims, providing other conditions within the scope of the respective independent claims, and giving the independent claim greater specificity.

All correspondence between the applicant's attorney and the Examiner during processing of the patent application is kept by the Patent Office in a "file wrapper" and is available to interested members of the public for review. During the processing, the applicant may be required to narrow a claim in order to get an allowance. That requirement results in a "file wrapper estoppel," which means that the applicant has been barred from extending the claim. This information is particularly important in the event of a subsequent challenge to the validity of the patent or in the case of an infringement.

OWNERSHIP RIGHTS AND INFRINGING ACTS

When no other parties are involved in a patent, the owner and inventor are one, and that person unambiguously acquires the monopoly rights granted by the patent. If there are two or more inventors, the law entitles each to an undivided interest with no accountability to the other, unless there is an agreement among them providing otherwise. However, the ownership rights of an inventor are often circumscribed to varying degrees, and sometimes absolutely, depending on the inventor's employment status, organizational affiliation, and/or the participation of others in the creation of the invention.

An inventor may lose ownership rights to the invention but does not lose identification as the inventor. The law requires that the actual inventor or inventors be named on a patent. It is illegal to omit legitimate inventors on a patent or to include anyone who has made no inventive contributions to the invention. Individuals or organizations that acquire ownership rights by virtue of their status as employers of inventors or sponsors of inventions are recognized as assignees on patents.

Conflict can occur during the patenting process or after a patent is issued if third parties appear who claim to have developed or commercialized the same invention, independently or otherwise. During the patenting process, this challenge can result in an interference proceeding. Once a patent has issued, the challenge would be taken up in the courts as a lawsuit challenging the validity of the patent. In the United States, deference is given to the first to invent. Most other countries give rights to the first to file a patent application.

Owning a patent means that third parties are in violation of the law if they make, use, or sell the patented invention without the owner's permission. Engaging in such acts without first obtaining the permission of the patent holder constitutes an infringement. A person is also liable as an infringer if he or she undertakes an activity that induces infringement. Such a person is liable as a contributory infringer, for example, if he or she knowingly makes and sells components for use in the infringement of a patent.

It is also an infringement to import into the United States a product of a U.S. patented process without the patent owner's permission. Also prohibited are the unauthorized selling of or the offer to sell a product arising from a patented process. Furthermore, the owner of a patent is entitled to legal recourse against anyone using his or her name or making reference to the patent in commerce with the intention of deceiving the public into thinking that such use or sale is authorized. It is also illegal to state for the purpose of misguiding buyers that a product or process sold is the subject of a patent or patent application if this is not the case.

If an infringement has occurred, the owner of the patent is entitled to legal remedies, provided that certain conditions have been met. The complaint must be timely. No damages will be awarded if the filing of a complaint is delayed for more than 6 years beyond the date of the infringement. The patented item must be marked with the word patent or the abbreviation pat. Such wording clearly puts the public on notice that the product is patented. In the absence of such marking, the patent holder must notify the infringer of the infringement. If the infringement continues, the patent owner will be entitled to recover damages.

The patent holder may be entitled to damages in an amount determined to be sufficient to compensate for the infringement. This amount may be tripled by the court. Considerations taken into account in determining damages include what might have been earned as reasonable royalty, as well as interest and court-determined costs.[4]

PATENTS AND ANTITRUST LAW

Under Section 1 of the Sherman Antitrust Act, it is unlawful to combine or conspire in restraint of trade; Section 2 prohibits monopolies and conspiracies or attempts to monopolize. Business practices in sales, leasing, pricing, and others that have the effect of substantially reducing competition or that tend to create a monopoly are proscribed under the Clayton Act, Section 3.

The prohibitions of these acts, however, do not generally apply to the patent holder, who holds a legal monopoly under patent law. The patent holder is allowed to engage in many otherwise illegal acts provided such acts do not exceed the legitimate entitlements of the patent grant. For example, an individual patent holder may exclude others from practicing his or her patent or may define territories for exclusive licenses. However, competing firms may not hold a pool of patents or engage in cross licensing. Such exclusionary practices and market sharing are prohibited by the Sherman Act. A single firm acquiring a portfolio of interrelated patents with the intention of monopolizing the market might also be in violation of the Sherman Act.

Sometimes a patent holder tries to require a licensee, under a provision of the license, to use a certain nonpatented product in combination with the product licensed. In effect, that uses the patent as a tying device. For instance, a licensee may be required to use a certain nonpatented oil in a patented machine. Such acts go beyond the prerogatives of the patent grant and could be construed as a misuse of the patent under the patent law or a violation of the Clayton Act if the effect is a material restraint of competition.[5]

More recently, the courts have made uncertain the rights of the patent holder to divide markets or to stipulate prices at which licensed products must be sold. While previous rulings have been supportive of grant-backs of improvement patents made by the licensee and have seen no distinction between acquiring an assignment by a monetary purchase and obtaining one as a consideration under a license, this matter is no longer clear under the law.[6] This increased uncertainty under the law raises the sense of risk in licensing and calls for greater scrutiny of the legal implications of related business arrangements.

Notes

[1] Peter D. Rosenberg, *Patent Law Basics* (New York: Clark, Boardman, Collagham, 1992), 2-1 to 9-18.

[2] Patent and Trademark Office, "Legal Analysis to Support Proposed Examination Guidelines for Computer-Implemented Inventions," *Official Gazette*, U.S. Department of Commerce (November 7, 1995): 3-10.

[3] Bobbi A. Brandon, "Deposit Requirements for Patent Purposes," in *Biotechnology Patent Conference Workbook* (Maryland: American Type Culture Collection, April 1989), 83-86.

[4] U.S. Department of Commerce - Patent and Trademark Office, *Patent Laws* Washington, D.C.: U.S. Government Printing Office, 1976): 44-48.

[5] A. D. Neale, *The Antitrust Laws of the U.S.A. 2nd Edition* (London: Cambridge at the University Press, 1970), 314-315.

[6] Douglas H. Ginsburg, "Antitrust, Uncertainty, and Technological Innovation," *The Antitrust Bulletin* (Winter 1979): 635-685.

PART II

Practice

Chapter 5

Defining Characteristics of the Market for Inventions

The monopoly gains that patents are intended to provide and that are potentially present under the protection of trade secret and patent law depend on the commercial success of the new product. In general, however, such gains are quickly eroded by market opportunities, as the forces of technological development and scientific progress in market economies bring competing products and processes into the marketplace. The commercial success of an invention thus depends to a large extent on the alternative choices available to buyers.

Competitiveness is a matter of survival in business, and licensing offers a unique opportunity for fresh approaches to market needs and cost efficiencies. The market for inventions therefore, is comprised of those parties who seek commercialization rights to inventions in order to develop and sell them as products or processes. A licensee may be an individual entrepreneur, venture capitalist or an established firm. Each faces a set of different situational circumstances. However, of interest to all is the viability of the new product/process candidate, and its ability to generate a favorable return on the investment required for its development.

LOCATION OF LICENSING OPPORTUNITIES

Would be licensees can learn about ideas for new products or process developments from a number of sources: publications of issued patents, journal articles, news releases, technology brokers, inventions management organizations, and private inventors. They can solicit new ideas, or new ideas may come to them unsolicited from owners seeking industrial sponsorship. Some corporations retain new-product-idea scouts.

Direct contacts with owners of inventions may be made through membership in licensing societies such as the Licensing Executives Society (LES) and the Association of University Technology Managers (AUTM). The growth in membership of these organizations attests to the increasing interest in licensing. Relationships arising from common life experiences (e.g., alumni and former working relationships), membership in scientific associations and participation in scientific conferences serve as invaluable avenues of contact.

Opportunities for middlemen have emerged, and technology brokers, as they are sometimes called, offer free services to the developers of new technology by listing inventions in their publications and databases. They make these listings available to industry representatives on a subscription basis.

Corporations solicit submissions of new product ideas by distributing copies of their annual reports and other material describing their interests in strategic product developments to the various sources of new technology. Many corporations have established technology acquisition departments to screen incoming ideas carefully before distributing them to the respective R&D units within their organizations. Patent departments within corporations ensure compliance with corporate policy and the law.

INDIVIDUAL ENTREPRENEURS

An entrepreneur is typically an individual with a strong interest in using a new idea to launch a new business. He or she may be a private individual or an employee such as a university faculty member who wishes to enter business. The entrepreneur must compete with established businesses in the field. These businesses often have advantages such as production and distribution facilities, advertising budgets, brand names, product acceptability, and access to capital. The challenge to the entrepreneur is to overcome these obstacles and to exploit the new opportunity successfully. To that end, the entrepreneur brings together people, capital, and the new idea.

However, while both existing firms and would-be businesses conduct searches for new product and process ideas, the owners of these ideas show a strong preference for dealing with established companies. This preference may be a function of awareness. While the prospective licensor is easily able to identify and solicit existing companies, the individual entrepreneurs seeking an invention on which to base a company are generally unknown.

When individual entrepreneurs do make contact with the owners of new ideas, they often are expected to demonstrate a greater ability to undertake the venture, both financially and technically, than is expected from an established firm. As a result, individual entrepreneurs rarely license new ideas from others, unless they have institutional sponsorship or the support of venture capitalists. The products and processes they bring to market are generally of their own making.

Development and Financial Considerations

The challenge entails a number of important phases. First, the entrepreneur must evaluate the commercial viability of the invention. Then the entrepreneur must formulate a strategy for its financing and development and must select a form of organization. These considerations must be carefully arranged in a business plan, which can then be used to raise capital and as a management guide.

Development of the business plan entails an assessment of potential demand for the new product or process, the nature of the competition, applicable government regulations, opportunities for patents, and the superiority of the contemplated product over competing products. Another important consideration is the potential of the new product idea to generate offshoot businesses and new product lines. High potential in this regard is an attribute of a seminal invention.

Other key considerations for the new business are the quality of the management team and the organizational structure. The choices made will be influenced by the developmental stage of the start-up, capitalization, and taxation. For organizational structure, the choices are sole proprietorship, partnership, and incorporation. While the sole proprietorship or partnership may be the preferred vehicle in the initial stages, in the long run, the corporation is the best structure for expansion, due to its ability to attract capital. It also limits ones personal financial liability.

The information developed above is important if the entrepreneur seeks financial backing. Would-be financial backers typically ask for a detailed description of the venture, its physical characteristics, and the basis for optimism regarding its viability. The latter includes the estimated profitability of the venture, its potential responsiveness to disruptions in conditions affecting salability, and fall-back positions. Short-term and long-term conditions must be expressed in dollar terms, as must projections of profitability and return on investment.

In seeking capital for a new firm, entrepreneurs are often confronted with the risk of diluting their ownership in the venture. Generally, the most desirable option is for the entrepreneur to retain as strong an equity position as possible, in order to ensure control of the undertaking. Accordingly, entrepreneurs should seek the most comfortable balance between equity and debt financing that will allow them to have the degree of control over management and ownership that they desire.

The issuance of stock, both common and preferred, raises capital but reduces the entrepreneur's equity in the business. However, stock is attractive in that it does not entail debt-servicing obligations, except that repayment may be required for redeemable preferred stock. At the other extreme, entrepreneurs can retain equity by taking out loans to raise the necessary capital. A drawback is that in debt financing, the payment of principle and interest, could be burdensome. Judicious use of convertible debt instruments offers the potential of achieving the best of both worlds.

Venture capitalists in particular make a high demand on equity and therefore constitute a capital source with significant dilution effects. Furthermore, venture capitalists prefer companies with the potential to generate returns many times the initial investment within a certain limited time period frequently 10 times the initial investment within 5 years. Because their stakes are high, venture capitalists often provide valuable management assistance to ensure the success of the venture. This approach also has serious implications for the independence of the entrepreneur.

Incubation

A form of business entry with low equity risk is offered by the leasing of facilities and equipment. Facilities and equipment costs combined with outlays for utilities, administration, market research, and even research and development pose a potentially severe drain on cash-flow. Leasing for the purpose of alleviating these costs may lead the entrepreneur to consider incubation for the start-up, or small business. Besides their lower costs and the valuable assistance they provide, incubators afford invaluable networking opportunities with other incubator companies.

Client companies receive space, facilities, and overhead services at below-market prices, often at less than half the going rates. They may also receive specialized services in business planning and management assistance in capital formation and marketing. If the venture is in the concept stage, sophisticated laboratory facilities accompanied by research and development support from faculty are often available in a university incubator. This support can result in the successful launch of an otherwise technologically immature invention.

Incubation opportunities are nationwide, and the National Business Incubation Association is a valuable information resource. Different types of incubators serve different objectives, thereby meeting the needs of varied business interests. There are incubators for the long-term, research-oriented project, and there are incubators with specific economic objectives, such as creating new jobs for displaced factory workers.

Creators of incubators include universities, private corporations, state or city governments, and venture capitalists. Often support is provided in return for a share in the equity of the new business. University incubators generally have the longest incubation term, since the companies they support are often involved in long-term research. Other incubator sponsors usually require that a client company be ready to transfer to the marketplace within two or three years. Incubators with broader geographic and economic objectives are more often sponsored by governments and private corporations.

To secure a site in an incubator, an entrepreneur must go through an applications procedure. Some of the requirements are similar to those imposed by banks for business loan applicants or by venture capitalists. An applicant is most likely to receive approval if the proposed company has a comprehensive business plan and a strong, well-credentialed management team. In addition, the sponsor may look at the personal financial commitment the entrepreneur is prepared to make as an indication of the entrepreneur's dedication to the success of the venture.

VENTURE CAPITALISTS

Venture capitalists play a unique role in the development of new products and processes. They have the ability to match funds with high-risk opportunities, and as such they are also agents of new business creation. Both existing firms and entrepreneurs seek them as a source of capital, as well as for the financial and technical expertise they are sometimes able to provide for the new start-up. Consequently, they join entrepreneurs and established firms as entities in search of promising product and process ideas.

Venture capitalists are those who raise and manage venture capital funds. They may be private individuals working independently, or they may be employees of subsidiaries of banks, insurance companies, or corporations that manage private or public funds. Venture capitalists look for promising business opportunities in which to invest. Generally, they expect to share in the equity of the venture and often expect to have some level of control of the operation of the business as well. The usual instruments for venture capital investments are common stock and preferred stock. The latter is convertible into common stock. They also use debt instruments.

Venture capitalists solicit venture capital funds from pension funds, corporations, insurance companies, endowments and foundations, and individuals and families. Foreign entities are also a source of funds. Institutional investors provide the bulk of funds available from both domestic and foreign sources. The magnitude of investment from institutions strongly affects the types of ventures preferred and the expectations for the timing and rate of return on investments.

Venture capital investment takes three major forms: start-up or seed capital, post-start-up capital, and special-situation capital. A high-technology invention that requires further development in order to reach market readiness is an example of a venture that would need start-up or seed capital. The start-up investment, therefore, is high risk, since it involves an untested product idea. Post-start-up capital is for the proven idea, the product or process that is already on the marketplace and that holds promise of high returns on the investment. Finally, special-situation capital is directed mainly toward acquiring control of mature companies, such as through a leveraged buy-out or the purchase of a subsidiary in divestiture.[1] Different forms of financing may be preferred for each situation, and within each depending on the level of risk taken.

The minimum amount of capital required for any one venture tends to be high. Early-stage start-ups can and do attract investors. However, the more mature ventures, the ones that require post-start-up and special-situation capital, offer faster returns on investments and tend to fair much better. Funds for these investments come from five distinct groups of venture capitalist funds. Largest are the megafunds. Others are mainstream funds, second-tier funds, niche funds, and corporate funds (financial and industrial funds).[2]

The megafunds are the largest funds and are invested globally. They tend to be private and not associated with corporations. Some form subgroups with specific technology interests. Mainstream funds, too, are mainly private and independent, but they include larger institutional Savings Bank Insurance Corporations (SBIC's). The SBIC's comprise most of the second-tier funds. These also include private and independent firms. The niche funds are all private and independent.

Niche funds tend to specialize in high-technology start-ups. Compared to the other funds, their resources are small. However, they do represent a revenue source for small start-ups. Second-tier funds have a more varied focus, but they also view start-ups as target investments. The focus of megafund, mainstream, and corporate funds is on later-stage ventures, leveraged buy-outs, and expansion financing. These investments offer quick and more secure returns. However, the large funds do occasionally support start-ups with particular promise.

Venture capitalist funds are typically constituted as partnerships, comprising limited partners and general partners. Limited partners contribute capital to the fund, but for tax and regulatory reasons, they must remain at arms length from the management of the funds. Limited partners include pension funds, institutional investors, and wealthy individuals. General partners are responsible for managing the fund and for its success. They raise the money, make the investment decisions, and operate the fund.

General partners are paid an annual fee for their services, usually based on the amount of the fund. They may also contribute capital to the fund, usually a very small percentage. They can receive up to 25 percent of the realized capital gains for as low a capital contribution as 1 percent. Unlike the limited partners, the general partners also bear a legal liability for the management of the fund.[3]

ESTABLISHED COMPANIES

The commercialization of inventions is important as both an offensive and a defensive business strategy. On the offensive are the entrepreneur and existing firms seeking market entry and acceptance for new products. On the defensive are the established companies, who must remain competitive to stay in business. The severity of the competition is evident in the following observation:

> . . . Even the biggest and best are aced out by their competitors on more than half their potential sales. . . . Eighty percent of new products vanish from the marketplace in two years. . . .[4]

Unless the firm or would-be firm develops an adequate research and development (R&D) endeavor, it must remain open to ideas emanating from third parties, such as universities, nonprofit and federal laboratories, and independent inventors, if it is to survive. Even for the company with an adequate R&D program, licensing-in broadens its options and reduces costs, thereby enabling the company to price its products more competitively and enhancing its probabilities of market success. Licensing-in accompanies corporate R&D and acquisition as the prime means for product development and diversification.

A corporation is more or less prepared to successfully undertake a new product venture depending on its awareness of market and technological realities; whether it keeps abreast of changing times. Sensitivity to consumer need, and technological applications commensurate with the times will determine its survival. These considerations entail an alertness to internal realities as well as the world outside.

Internally, the company must be aware of its financial and technical ability to engage in product innovation; externally, to changing technological, competitive and demand patterns. It is important to keep apprised of both internal basic health and growth needs and to remain alert to potentially inimical market and technological developments.

Licensing-in as a means to corporate growth and development is likely to increase as R&D costs escalate and as product life cycles contract. This new product option is relevant for both innovation in existing lines and introduction of new lines. To remain competitive, products must be both competitively priced and at least as desirable as the alternatives if price is to work in the company's favor. Licensing-in reduces the costs at which this may be achieved. The bulk of the cost of acquiring commercial rights to an invention are payable as royalties when sales occur.[5] R&D costs are minimized; the lead time for market introduction of the new product is reduced as is the risk of market failure.

However, establishing a need and justification for using an outside idea are not the only requirements. Corporate culture, departmental idiosyncrasies and the personalities of powerful individuals in the corporation play a vital role in the success of a project. Technology to be licensed must be compatible with the culture of the corporation, and a managerial consensus is imperative. A prospective idea must fit with the corporate strategic plan, and acceptance of the idea must not be dependent on the support of only one corporation official.

Dependence at first will be on the initial contact person, but the new product candidate must eventually pass multiple decision points. In this regard licensors should remember that territoriality is a common trait in corporate life. For example, an offense to the R&D department could interrupt the progress of an idea originating in the Business Development or Marketing department. In addition, many companies are afflicted with what is commonly know as the "not invented here" (NIH) syndrome. These companies generally prefer internally generated ideas, as long as the R&D Department is staffed and equipped to meet strategic goals.

Progress of an idea toward selection as a licensing candidate may also be interrupted by the idiosyncrasies of certain professional types. Engineers or scientists might look for technological challenge, disregarding production feasibility, salability or profitability. The manufacturing people might look only at production feasibility, marketing people at salability and the financial people at the bottom line of cost and profit. Resolution of these disparate needs requires both cooperation, and a mentor for the idea who is able to develop the necessary teamwork required for the success of the candidate product.

The conditions under which companies are prepared to consider unsolicited ideas also vary. In one study a survey was taken of 1,200 companies for attitudes concerning external submissions. Usable responses numbered 243, with results as follows: (a) 112 examined ideas after receiving a signed waiver from the submitter; (b) 31 examined patented ideas only; (c) 9 rejected all unsolicited ideas without examining them, and informed the submitters of this policy; (d) 8 ignored all unsolicited ideas; (e) 11 checked the "Other" category of the survey.

A total greater than 243 is obtained from the above numbers. This is due to multiple checks by respondents. Most numerous are the checks of combinations of (a) and (b). A requirement of a waiver in the absence of a patent is the explanation from some. Others checked (a) and/or (b), first screening for promising ideas before applying these responses. The (a) and/or (b) responses reflect a concern in industry of compromising rights in technologies that might be under development in its laboratories.[6]

DETERMINING COMMERCIAL VIABILITY

Companies, entrepreneurs, and venture capitalists use different criteria in selecting new product ideas for commercial development. Such criteria are relevant to both home grown and externally sourced ideas. Another study identified 86 screening items, which the authors reduce to a factor representation of 11 screening dimensions. Dominant among these are, financial potential, corporate, technological, and production synergy and differential advantage, in that order.

Financial potential, the most important screening dimension, includes sales expectations, profit potential, and likelihood of success. The synergy dimension relates to the company's ability to develop and manage the new product within existing structures of organization and know-how; in other words, the inventions fit into the company's current business, sales and distribution channels, technological facility; production skills; and resources. Differential advantage refers to the potential of the product idea to become a first-on-the-market technological breakthrough.[7]

The first dimension, financial potential embodies commercial success. Companies exist to create profits for their owners and investors. They do this by producing products that customers will buy. Market acceptance provides the opportunity to make a profit and grow. Many methods for measuring financial potential exist. One of the more important of these is a method for estimating the rate of return on the investment made in a new product. This method, also known as the discounted cash flow or investors method is the traditional method for determining the present value of an annuity. Alternatively, the method can be used to calculate the net present value of a proposed project. The formula is,

$$C = \frac{R}{(1+i)^n} + \frac{A}{(1+i)^n}$$

Where,

C	=	Investment outlay,
R	=	Expected annual cash inflow (before depreciation and after taxes)
i	=	Rate of return,
n	=	Expected life of the project,
A	=	The recoverable asset value at the end of the project (assumed to be zero for the present purposes).

With C, R, A, and n given, an (i) that equates the discounted cash flow with the investment outlay can be obtained. This rate is the floor below which the project is not feasible. Candidates must demonstrate higher returns. Alternatively, the rate of return (i) is the highest rate that can be paid for funds to finance a project without sustaining an economic loss.

Table 1 is illustrative. Assume an estimated cost of investment of $758,000 and the indicated cash inflow. At 10 percent, the present value of the cash inflow is equal to the required investment outlay. Two interpretations might be given to this data, depending on whether the company must borrow money for the project or if it already has the funds.

Table 1
Calculation of Present Value of an Investment

		Present Value at 10% of	
Year	Cash Inflow	$1	Cash Inflow
1	$200,000	.9090	$182,000
2	200,000	.8264	165,000
3	200,000	.7513	150,000
4	200,000	.6830	137,000
5	200,000	.6209	124,000
Totals	$1,000,000		$758,000

The 10 percent rate is crucial for both cases. In the former instance, borrowing at higher rates would be uneconomical. Rates greater than 10 percent could be sustained only by higher cash inflows. Regarding the latter, the funds might be directed to alternative investment opportunities yielding returns greater than 10 percent. In economic terms, the best rate lost as a result of taking the 10 percent option is the opportunity cost of the investment.

More generally, with (i) as the denominator and the rates of competing projects as the numerator, projects may be arrayed for screening purposes. Economically viable projects are those with ratios equal to or greater than "1." Also, with (i) set at the company's cost of capital, a project is considered acceptable if the net present value of the discounted cash flow equals or exceeds the required investment outlay. Setting the required investment for each project as the denominator for its present value estimate, all projects with ratios exceeding "1" would be considered satisfactory, and the greater the excess, the more desirable the project.

Notes

[1] Gilbert N. Dorland and John Van Der Wal, *The Business Idea* (New York: Van Nostrand Reinhold Co., 1978), 68.

[2] William D. Bygrave and Jeffry A. Timmons, *Venture Capital at the Crossroads*. Boston, Massachusetts: Harvard Business School Press, 1992), 52-58.

[3] Ibid., 11-12.

[4] Harvey Mackay, "Mackay on Business: Sooner or later, the day will come when we have to learn to be losers," the Albany New York *Times Union* (May 21, 1995): B-1.

[5] John W. Morehead, *Finding and Licensing New Products & Technology from the U.S.A.* (Elk Grove Village, Illinois: Technology Search International, Inc., 1982), 21.

[6] G. G. Udell and K. G. Baker, *PIES-11 Manual for Innovation Evaluation*. Madison, Wisconsin: University of Wisconsin Extension, 1984), 163-166.

[7] R. G. Cooper and U. de Bretani, "Criteria for Screening New Industrial Products," *Industrial Marketing Management*. Vol. 13 (1984): 149-155.

Chapter 6

Locating an Industrial Sponsor

The ability to commercialize inventions is not available to all who develop them. Due to their status as either public or nonprofit entities, government laboratories, universities and research institutes generally rely on the industrial sector to bring about market entry for the new technologies that they create. Private individuals, as independent inventors also often take the licensing route for commercialization of their ideas. The business community provides the vehicle for market entry of inventions owned by these parties.

Companies may be identified as to whether or not they are likely to be good licensees. The candidates vary also in their starting positions. There is the start-up or spin-off company. This kind of company may be headed by an individual entrepreneur or venture capitalist entrepreneur. However, it is the population of established firms that is most familiar to marketers of inventions. The challenge is how to sort out from this very large population those that should be approached as potential licensees. This challenge must be faced by both the marketer of an individual invention and by those managing invention portfolios for institutional owners of new technologies.

CONSIDERATIONS IN SELECTING A LICENSEE

Most inventions face severe competition for licensees. Thus, licensors often accept the first company willing to negotiate a license. The excitement of finding a licensee often overshadows a licensor's inclination to evaluate the licensee's ability to introduce the new product into the market. Considering the high rate of market failure for new products and the low number of licenses that actually generate royalties, the licensor is advised to proceed with caution in selecting the one to commercialize the invention.

The approach taken by companies in choosing which ideas to develop into commercial products is instructive. Among prospective licensee, the conventional wisdom is to examine carefully many promising opportunities and their fit in the market before deciding on the candidate of choice. Prospective licensors should be equally careful to examine the market conditions surrounding an invention and the relative capabilities of interested licensees to make the invention a commercial success before entering into a license.

A well planned marketing program will increase the probability of locating the best and most capable licensee. On the other hand, if the invention is not accepted as a new product or process candidate, the inventor has the option to use the feedback from the industrial evaluation to either modify the technology so as to make it more licensable, or to develop a more market-oriented direction of research, should he or she so choose. Company-specific technological needs may remain unknown to inventors who rely on the opinion of "experts," especially those who advise them to abandon inventions without first having direct contact with would-be licensing candidates.

If the invention proves to be commercially viable and a number of companies show interest in licensing, the relative strength of the candidate's marketing ability should be a major consideration in determining which one among them to choose as a licensee. Criteria in this regard have been suggested.

Criteria for assessing a firm's ability to launch a new product were established in a 1989 joint study by the Small Business Development Center of New York State and the Research Foundation of State University of New York. An initial list of 23 criteria was reduced to seven key informational predictors and ranked as follows:

1. Percent of sales from products introduced in the last 5 years;
2. Number of new products, or innovations to existing products, introduced in the last 5 years;
3. Number of new product or process technologies licensed in or acquired in the last 5 years;
4. Percent of sales used for new product development;
5. Return on investment;
6. Number of employees in engineering and/or R&D;
7. Average age of manufacturing equipment and processes.

This ranking was developed from weights that participants were asked to assign to the criteria. The participants were "experts," persons with executive-level experience in firms introducing new products, and liaison persons involved directly in technology transfer as marketing and licensing professionals.[1] Persons marketing inventions could apply similar criteria to the companies in the particular industry of the invention under consideration as part of the licensee selection process.

Choosing the right licensee(s) is also addressed in the inventions licensing literature. One study cites research and development, manufacturing and sourcing, and sales and marketing as skills that can be readily evaluated for comparative purposes. The importance of these skills for success varies by industry, and their relative significance is expressed in their respective cost relationship to total expenses.

Of particular note is the importance of R&D for early-stage technologies and for companies engaged in global competition. Companies with a competitive edge in manufacturing and sourcing skills lead in industries with high direct costs relative to sales (low gross margins). Sales and marketing as a criterion for effectiveness is most important in industries with a large customer base. It is also important for products with complicated technical merit.[2]

Another important consideration in determining company selection is the stage of development. The invention may still be at the concept stage; it may be in process towards reduction to practice; or it may already be reduced to practice and be demonstrating utility. The first two stages represent partially developed technologies. From the commercial view, these technologies are embryonic, high-risk ventures. Inventions in the third stage have been substantially advanced towards being market ready. These stages of development provide opportunities for "high-risk" investors and for "risk-averse" investors, respectively.

Businesses respond differently towards the various stages of development of an invention. The high-risk takers are most likely to be receptive to embryonic, high-risk inventions. An interest in the potential of the technology to create wealth is their guiding principal. The wealth in this case is differentiated from that earned in speculative transactions such as in the stock or real estate markets, where the gain involves a transfer rather than a creation of wealth. Companies under the control of pension funds and stock speculators tend to be driven more by the promise of short-term gain, and therefore tend to be less supportive of risky, long-term commitments.

Since embryonic inventions generally require a long-term commitment, firms that are controlled directly by their owners tend to be the most receptive to taking a chance on promising new high-risk technologies. These are parties who might be interested in developing a family fortune. Larger firms in which a party or family holds enough shares to influence management decisions may also be interested in the high-risk invention, and its wealth-creating potential. Firms with little to lose in capital investments dedicated to other products are also more likely to undertake a new investment. Small firms, or start-ups, are thus considered good candidates for the high-risk inventions.[3]

However, most companies, whether risk-averse or not, will be attracted to inventions that hold market promise and that have been substantially advanced toward market readiness. The future of such inventions is less uncertain. Also, in general, it is the large, established firms that constitute the preferred first target for many licensors. Such targeting is not without results. Even companies not under the control of owners will encourage "intrepreneurship" in order to remain competitive. In this latter instance, people often emerge who have a personal stake in the success of the venture.

CONFIDENTIALITY

In the absence of an issued patent or related information already in the public domain, information concerning the invention should be disclosed only in the form of a nonconfidential description. The response rate of licensing candidates will vary, depending on the field of the invention, and the marketer should not be discouraged by the seeming low rate of interest. Most responses will be outright declinations. When a company does indicate interest and asks for more details, the marketer should have it execute a technology evaluation agreement before divulging any details of the invention.

The confidentiality agreement is an example of a technology evaluation agreement. Under this agreements a company promises not to use the information it receives for commercial purposes without first entering into a license agreement. Another is the testing agreement, under which a company may be given a chemical compound or a prototype of the invention to determine if it meets its market needs. Under a screening agreement, a company receives compounds for evaluation, obtaining licensing rights for those that show the desired activity. An option agreement provides a company with an exclusive period of time to determine its commercial interest, in exchange for compensation.

A common feature of the technology evaluation agreements is the confidentiality provision, and the promise they hold for a possible license. Appendix A provides sample agreements.[4] By requiring confidentiality the agreements protect patent rights in the absence of issued patents, and minimize the risk of unauthorized appropriation of the technology by the receiving company. However, the agreements do not guarantee a license.

NEW COMPANY CREATION

The noncommmercial institutions are now actively involved in both company start-ups and in the search for commercial sponsors for the start-up companies. Many of the nonprofits encourage employee entrepreneurship. However, because the inventions are generally owned by the institution, the employee must acquire a license in order to start a company. Often, the license agreement results in an equity position for the institution. This form of technology transfer frequently represents a viable alternative to the traditional licensing of established companies.

For those institutions that encourages employee entrepreneurship, the success of this licensing mode will necessarily be influenced by the ability of these engineer/scientists to launch a new business. A factor in their success will be the inducements they receive from their employer institution.

Institutionally sponsored entrepreneurs have a potential advantage over competitor entrepreneurs who have no institutional affiliation as they seek new product ideas or attempt to start a business. Institutionally affiliated technology entrepreneurs also have an advantage over independent entrepreneurs in their ability to attract venture capital, since venture capitalists see the research institution as a possible agent for continuing product enhancement, particularly in high technology lines.

In funding a new start-up, a venture capitalist acts primarily as an investor. Venture capitalists also act as entrepreneurs, assembling their own management teams for purposes of starting and managing new businesses. In this role, they are available as licensees, and they are approached with product ideas (inventions) by both independent inventors and institutions. Those seeking venture capitalists as licensees have a number of sources to investigate. One is the INTERNET; another the local library. Publications such as the *Venture Capital Journal*, and *Pratt's Guide to Venture Capital Sources* are also good sources for leads.

Representatives from venture capital firms participate in business and technology trade shows and conferences that showcase promising new business opportunities. Such events enable potential licensors and licensees to make direct contact. Help can also be obtained from individuals knowledgeable about the whereabouts and interests of venture capitalists in the licensing business. These avenues toward venture capital contact are available to both independent inventors and licensing professionals who represent institutional inventions.

Besides soliciting venture capitalists to become licensees, institutions solicit venture capital funds as part of the resources they create for new start-ups by their employees. As noted previously, these institutions may also create incubators where would-be firms, headed either by entrepreneurial employees or by independent entrepreneurs are nursed until ready for the real world of market competition. Support from venture capital may be used here as well. The licenses that result generally include equity positions for both the institution and the venture capitalist.

Institutions may also create venture-affiliated funds to advance commercialization of their inventions. The targeted technologies typically are those with promise to fill a technology gap. Favored among the technologies are the seminal inventions, which have the potential for wide applications and many product lines. From the experience of many institutions with this avenue of technology transfer, a number of models have emerged. The Harvard model is illustrative.

At Harvard, the venture-affiliated fund concept was pursued by its Medical School. Harvard launched a traditional independent venture-capital partnership, based on a target fund of $35 million. The money was raised by its Medical School, with the purpose of advancing the commercialization of inventions made at Harvard and its hospitals. The partnership was named Medical Science Partners. It consisted of a general or managing partner and the investors or limited partners. Harvard created a not-for-profit entity to represent it in the fund as a passive, limited partner.

The fund paid a standard management fee to support its operations. Inventions developed by the university received 85 percent of the capital raised. Eighty percent of the returns went to the limited partners, 10 percent to the general partners and 10 percent to Harvard. Harvard dedicated its receipts to research and teaching. The fund raising ended in 1990 with a yield of $36 million. Fifteen investors were represented: one-third from the United States, one-third from Europe, and one-third from Japan.[5] Aside from the problem of fund raising, a major obstacle to be overcome in creating a venture-affiliated fund involves conflicting interests. Within the university, conflicts of interest can occur among faculty and between faculty and administrators. Academic and business investors may have conflicting interests. Investors can include individuals, institutions, and corporations; they too can be in conflict. Balance among these various concerns is crucial for the success of a venture-affiliated fund.

ESTABLISHED COMPANIES

The most common sponsors sought for commercialization purposes are the established manufacturers. They number in the millions and are located throughout the world. For the inventor seeking a licensee, it is important to know where and how to look for an appropriate sponsor among them. To locate a licensee among established firms, the technology marketer must be knowledgeable about the particular industry to which the new technology applies, about related products on the market, about the relative advantages of the invention, and about the potential of the invention to make a profit. Potential profitability is a prime selection criterion for a manufacturer.

The technology marketer should target initially those companies that possess suitable manufacturing and distribution facilities and a capability to launch the kind of new technology the invention represents. Companies that produce products similar and/or related to the new invention demonstrate that capability, and these companies are the most likely to have the capability to make the invention a success on the market.

Companies identified as potential licensees are often multi-product firms. These often have affiliated interests that are constituted as separate entities, such as branch operations, subsidiaries, or divisions, each with its own departments and product-line champions existing under the umbrella of the parent company. The technology marketer may find that a number of departments become involved with the external submission, each with its own particular mission in mind.

The departments commonly involved in screening incoming ideas are marketing, legal/patent, engineering, and research and development.[6] To these may be added the technology acquisition department, which some firms create specifically for this purpose. Effective marketing may require the technology marketer to work patiently with each company contacted and through various corporate structures and departmental contact points. The goal is to locate within each company those persons who are ready to argue that licensing the new idea will contribute to corporate growth.

The exposure possible for the invention depends on the budget of the marketer. For a lucky few, the quest for an industrial sponsor may produce quick positive results. Most inventions, however, rely on an extensive targeted search of primary and secondary sources for potential sponsors. Industrial manuals and directories are examples of secondary sources. With the advent of the Internet, key-word searches provide another option.

The better secondary sources use classification schemes under which corporations are grouped by the products they produce. This is particularly helpful in identifying companies with a possible interest in the invention. Most prevalent of the classification schemes used is the standard industrial classification code (SIC code).

The SIC code is used in industrial manuals and State directories of manufacturers, in government reports on industrial performance, and frequently in industry publications. Many of these publications are held in local libraries, and marketers can readily identify potential industries and companies for the new technology through publications using this system.

Among the commonly known industry publications are Dunn & Bradstreet's *Million Dollar Directory*, the *Thomas Register*, and the *Corporate Technology Directory* (Corptech). In addition to providing SIC codes, the latter have their own classification schemes, identifying companies at highly specific product levels of detail, thereby enabling fine targeting of companies. The Thomas Register has product lines listed alphabetically by which companies can be identified. The Corptech reference comes with a technology index by which the seeker can be guided. It also gives the names of key corporate executives and their titles.

In contacting a company, the licensor should personalize the communications and address them to a named individual. Ideally, the communication should go to someone with responsibility for new-product acquisition/development at the company. The communication should be accompanied by a brief, nonconfidential description of the invention, highlighting its advantages over existing competitive products/processes, and indicating whether or not it is patented. If it is patented, a copy of the patent should be included with the mailing.

There are other ways of making the invention known to industry, such as *new*- technology catalogues, newsletters, and commercial databases. These methods are less expensive but also generally less effective than direct marketing. While listing the invention in databases and in technology transfer bulletins would seem to be a logical recourse for reaching the most companies possible, one study showed that such listings seldom generated inquiries in the vast majority of cases.

The study examined 10 cases in the biotechnology and measuring, analyzing, and controlling instruments (MACI) fields. Marketing of these technologies involved all of the techniques indicated: direct-contact mail, targeted brochure distributions, and listing by commercial technology agents in computer databases and newsletters. A record was kept of companies dealt with directly for individual inventions. A total of 524 direct contacts were made for all the inventions, ranging from 20 to 100 contacts per invention.

The sample generated 70 confidentiality agreements, ranging from 5 to 10 per invention. Of the 70 confidentiality agreements, 54 resulted from direct, individualized mailings, 14 were traceable to brochure-related inquiries, and only 2 appeared to originate from contacts through commercial databases and technology newsletters.[7] Therefore, by far the best approach is the individualized approach. New technology brochures featuring similar invention ranks a distant second.

This experience affirms the common belief that a brief, well targeted nonconfidential description of the invention is the best way to inform industry about a licensing opportunity. If there is interest, the best way to follow up is under a confidentiality agreement, with a thorough disclosure of the available technology so as decision makers fully understand the opportunity. The following may be some of the reasons why company executives prefer the direct approach:

1. Professional etiquette requires corporate contacts to respond to personalized mailings, giving priority to direct mailings about a particular invention.
2. Many corporations are recipients of a large number of targeted, unsolicited, new- product ideas. Time may simply not permit and/or conditions may not warrant the additional expenditure of time to look for new ideas by wading through ostensibly equally meritorious ideas in computer databases and technology newsletters.

Competition among inventions exists not only at the invention level but also at the institutional level. Some inventions are helped by the stature of the institution where they were invented, by the ability of that institution to provide additional support, and/or by new developments on a licensed technology. By packaging similar technologies, a relatively unknown institution can convey to the industrial sector the scope and depth of its technological capability. Also, a company and/or venture capitalist might be attracted by the potential for a strong market position offered by a package of related technologies.

The benefits of presenting an invention to potential licensees are not restricted to the prospect of a license and subsequent revenues. The inventor and the invention can also benefit from continued exposure to the scrutiny of the industrial target market. By taking advantage of industry evaluations, detailed industry critiques, and the inventor's own heightened awareness of the relative financial, technical, and market limitations of target companies, the inventor can revise the invention and/or marketing strategies to improve its probability of success. The inventor should expect to receive a detailed evaluation report from any company that requests privileged information from the inventor under a confidentiality agreement or other such evaluation instrument.

MANAGING THE MARKETING OF MULTIPLE INVENTIONS

An adequately staffed and financed technology transfer operation has a variety of primary and secondary sources of industrial interest available to it. Opportunities for primary contacts exist through memberships in technology transfer societies such as the Licensing Executives Society (LES) and the Association of University Technology Managers (AUTM).

One goal of such organizations is to promote opportunities for personal contacts between potential licensors and licensees and with colleagues at other institutions. They offer opportunities to get to know the face at the other end of the telephone. Members can discuss mutual interests and exchange brochures on a person-to-person basis at association meetings and conferences. These communications are beneficial for immediate objectives as well as for long-range working relationships.

Corporate interests can be tracked by analyzing their responses to previously submitted ideas. Marketers can then target those companies that demonstrate the greatest receptivity to their new technologies. Institutions with a large number of inventor clientele, such as universities and government laboratories, are beginning to use computer technology to generate finely directed mailing lists for new disclosures based on relationships and prior experience with target markets.

To prioritize selections, responses may be scored. Under this scheme, companies with agreements on other inventions in the subject technology receive the highest score. These are companies demonstrating a seriousness about dedicating time to evaluating the external submission for licensing purposes.

If there is interest in the submitted idea, these companies are often willing to sign the necessary agreements in order to get a full appreciation of the technology before entering license agreements. The presence or absence of such agreements on previous company contacts or feedback on such arrangements can help the marketer rate the potential interest of companies in the target market and thereby better direct future inventions in the subject technology.

Use of a technology coding and company rating scheme may proceed in this manner. Inventions promising a cure for cancer may be given an SIC code at the 7-digit level (e.g. 2834-041). Now, consider a technology transfer office (TTO) with three inventions in its portfolio that claim to be cures for cancer and on which the TTO has developed a marketing history. Each invention has accumulated a response record based on inquiries initiated by listings in technology brochure distributions, listings in commercial databases and publications, direct mailings, etc. . . . Each of these inquiries has been rated according to the degree of interest eventually expressed by company contacts after the companies complete evaluations.

A new invention coming in with a promise to cure cancer will be coded 2834-041, accordingly. Interest solicitation begins with the TTO's internal database. A report is run by SIC code, listing all the companies under 2834-041 with which the TTO has had dealings. The companies are ranked according to potential interest, and the newcomer is prepared for mailing first to those companies most likely to respond positively, based on prior experience.

However, other considerations should also be taken into account in order to enjoy the full benefit of prior experience. The rankings should not be substituted for professional judgment. Their purpose is to facilitate screening by ordering the data. This becomes increasingly useful with the growth of the history file and where contacts under some SIC codes may number in the hundreds.

Examining listings based on agreements record may not always be helpful since marketing often fails to produce such arrangements. For most, the TTO will need to examine the history files for such phenomena as evidence of the NIH syndrome, companies that tend not to respond to correspondence, former contacts that may have moved out of the SIC category of the invention. A company that was previously licensed in the subject technology might not be ready to take on a newcomer but may wish to continue receiving information on new developments. Another company may sign a confidentiality agreement for details on every new technology in this field without ever intending to take a license, but to augment its technology library and to keep abreast of emerging technologies.

Occasionally, companies will make "shotgun" inquiries regarding inventions listed in technology brochures. Comprehensive follow-up material is provided under multiple confidentiality agreements and these companies are never heard from again. Companies that do not exhibit serious interest in licensing but nonetheless continue to encourage solicitations should be identified and approached to determine if there is fault in the marketer's methods of follow-up, if there is a weakness in an understanding of the company's strategic interests by the marketer, or if the failing is in the competitiveness of the technology.

Notes

[1] Small Business Development Center and The Research Foundation of State University of New York. *Workshop Concerning a Firm's Ability to Launch a New Product,* held at the Rockefeller College in Albany, New York, in 1989, ____.

[2] Mark Speers, Paula Ness Speers and Lisa Pisano, "Choosing the Right Licensee(s)," in *AUTM Technology Transfer Practice Manual.* Connecticut: Association of University Technology Managers, Inc.: Connecticut, 1993. Part IV, Chapter 3, 1-17.

[3] John T. Preston, "Key Problems in Commercializing Technology in the U.S.," Presented as Testimony Before the Energy Subcommittee of the House Space Science and Technology Committee (March 1993). Copies of Material Distributed for a 1995 Presentation at the Rensselaer Polytechnic Institute, Troy, New York, 3.

[4] The Research Foundation of State University of New York. The documents in Appendix B are samples from working documents in the Foundation's Technology Transfer Office, and some of these have been modified by the author for this work. The Foundation does not claim to endorse their application for this work, and the author takes full responsibility for their use.

[5] Stephen H. Atkinson, "University-Affiliated Venture Capital Funds," *Health Affairs* I (Summer 1994): 166-167.

[6] G. G. Udell and K. G. Baker, *PIES-11 Manual for Innovation Evaluation* (Madison, Wisconsin: University of Wisconsin, Extension 1984), 163-166.

[7] Albert E. Muir, "Managing Inventions Marketing," *les Nouvelles (Journal of the Licensing Executives Society)*, Vol. XXV, No. 4 (1990): 188-190.

Chapter 7

Determining the Value of an Invention

Considerations of value must be resolved before a license agreement can be concluded. Value means how much the licensor is entitled to as compensation from the licensee in return for licensing rights. Compensation, or royalty is income for the licensor, and a cost for the licensee. Therefore, one will want more, and the other less as consideration in the agreement. During the negotiation process, both parties strive to arrive at mutually agreeable royalty terms.

Concepts of fairness and their resolution have been addressed by the courts. Payments may take different forms and methodologies exist enabling fees or royalty rates to be expressed in their equivalent dollar amounts thereby enabling the parties to make fair trade-offs between these alternative forms of payment. The parties may also wish to view an array of variables in their negotiations and these may be expressed in either dollars, importance weights, or combinations thereof to arrive at royalty rates. Most helpful is a model that enables the parties to calculate the impact on royalty of changes in valuation of the variables. Economic models also show that royalty rates and compensation by fixed dollar payments have differing effects on the pricing behavior licensees. This has implications for output and earnings.

FAIRNESS IN DETERMINING VALUE

The determination of royalty entails considerations of equity and fairness. Many issues of varying importance to the parties are involved. Before engaging in the negotiations, each party must decide, usually on its own, its own preference for the calculation method for the royalty, the form payments should take, and the frequency of payments.

Concerning fairness, two considerations are relevant. One relates to the disparate treatment of development costs in royalty determination, and the other to the degree of participation of the parties in the success of the commercialized product. The expectation is that the licensing agreement will arrive at a fair distribution of gains between the two parties, that is, the more one party contributes to the development of the licensed product, the more that party will gain.

Generally, however, the licensor's costs of development are treated as sunk costs and do not enter the royalty formula. For the licensee, the fixed costs incurred in developing the invention to commercial readiness comprises its investment in the new product. These costs, and the returns they generate, constitute the financial concept of return on investment, an important criterion in the decision to launch a new product.

Whether or not development costs are considered, the sunk costs of the licensor do relieve the licensee of costs that would otherwise be necessary, thereby reducing the required level of investment and the risk of developing the new product. If the sunk costs had been borne by the licensee, the expense would be treated as a legitimate component of recoverable cost. This matter has been addressed by the courts, and R&D costs have been taken into account in the determination of reasonable royalty.[1]

The degree of participation in the success of the commercialized product is the more common reference for calculating royalty and fairness. Among the possible forms, the running royalty is the most popular and usually gives the licensor a percentage of net sales. This sharing of success, achieved through a distribution of the profits, is considered a logical indicator of fairness. Experiences indicate that royalties range from 15 to 50 percent for the licensor. A share of 25 percent of profits is commonly accepted as a reasonable royalty.

However, using profits as a basis of royalty payments may be problematic, because of the different accounting ways that costs can be represented. Preference is generally for a royalty to be calculated as some percentage of net sales; 5 percent is reasonable, for companies making profit in the 10 percent to 20 percent range. A rate of 5 percent of net sales, on the 25 percent of profitability commonly accepted as a reasonable participation in profits, would represent a profit-to-net-sales ratio for new products of 20 percent. For example, net sales of $1,000 yield a profit equal to $200. Twenty-five percent of $200 is $50, the amount paid as royalty. This $50 represents 5 percent of the net sales of $1000.

The courts have offered varying rates for fairness depending on the circumstances of the case. For the use of a synthesized manufactured vitamin, 10 percent of sales was judged fair. In another case, the court awarded 2 percent for a device for fire extinguishers. The rates awarded by the courts are percentages of the selling prices of infringing products. They range from 2 to 20 percent and are indicative of valuations that licensors and licensees need to be mindful of under similar conditions.[2]

TYPES AND CALCULATIONS OF REMUNERATION

While percentage of net sales appears to be the most common way to calculate royalty payments, it also may take many forms. The percentage rate might be constant, it might decline beyond certain base levels as sales rise (providing an incentive to the licensee to promote sales), or it might increase at higher volumes. Lower rates at lower sales help the company get established in the market. While both licensor and licensee gain when a product is profitable, a rate on sales penalizes the licensee if the sale is generating a loss. A running royalty is advantageous to the licensor in that income is earned even if there are no profits, and the rate of return is automatically adjusted for the effects of inflation.

Royalty rate-based payments may be accompanied by an up-front payment, also called a license issue fee, and/or a minimum annual royalty payment. The up-front payment is perceived as a demonstration of commitment by the licensee. Also, since rate payments usually lag behind the agreement, often by several years, the up-front payment and the minimum payments provide immediate returns to the licensor with which to offset development and licensing costs. They also serve as an inducement for the licensee to bring the licensed technology to market quickly.

A variant of the minimum payment is a periodic lump-sum payment unaccompanied by a royalty rate-based calculation. This form of payment is often used when calculating a running royalty is impractical, as in the case of a process license or when the technology is a small component in a complicated system. Alternatively, the licensor may receive a fixed sum for every unit sold. Tying such payments to some economic index ensures protection against the corroding effects of inflation.

Royalty may also be paid on proceeds the licensee receives from its sublicensees. If the licensee is simply a conduit to the sublicensee, such rates can range as high as 95 percent. Another means of compensation is stocks, or equity in the licensee, thereby giving the licensor both participation in profits and increased control in the licensed business.

Publications are available showing royalty arrangements in different industries. One study illustrates low and high royalty rates achieved in selected industries. The figures reflect patterns in licensing agreements of U.S. corporations relating to patents, trade secrets, and know-how. Questionnaires were sent to 150 randomly selected corporations, resulting in 37 usable responses from companies with sales exceeding $500 million.

The data showed a strong preference for straight royalties. Paid-up licenses were next, then down payments with royalties. Patent licenses had a low of 0.2 percent and a high of 35 percent of sales. The low for trade secret licenses was also 0.2 percent and the high for these was 15 percent. The identified industries were, biological, chemical, electrical and mechanical, and the high occurred under mechanical licenses.[3]

The great variability evident in industry rates and in the rulings of the courts is attributable to a number of considerations, among which are variable profitability (affecting rates based on sales), varying degrees of participation in the development of the licensed product (affecting relative participation in profitability), and varying perceptions of fairness.

Costs can be used as a basis for estimating the minimum royalty the licensor should receive and the maximum royalty a licensee would be expected to pay. However, if royalty rates are to be the guides for remuneration, they should undergo certain analyses to match returns (royalty) and costs. Such analyses examine the expected performance of the licensed technology as a product on the marketplace relative to the investment in developing it. For the licensor, the investment is past; it is the sunk costs of creation. However, the method of present value used in Chapter 5 for licensee calculations is applicable here as well. Table 2 is reproduced from that chapter for illustrative purposes.

Table 2
Calculation of the Present Value of a Royalty

Year	Expected Royalty	Present Value at 10% of $1.00	Present Value at 10% of Amount
1	$200,000	.9090	$182,000
2	200,000	.8264	165,000
3	200,000	.7513	150,000
4	200,000	.6830	137,000
	200,000	.6209	124,000
Totals	$1,000,000		$758,000

The formula R/(1+r)n is applied for the present value of $1.00, with $1.00 as the numerator in each of the five years. Table 2 is illustrative. Multiplying Expected Royalty by the result in the "$1.00" column gives the present value of $200,000 in each of these years. The Amount is rounded off to the nearest thousand dollars. Assume sunk costs of $758,000 and royalty at 5 percent of net sales yielding the royalty indicated. At 10 percent, (i.e., market interest rate), the present value of the royalty amounts to $758,000 and just pays for the sunk costs.

In the above example, the present value method determined the "fixed-sum" equivalents of a running royalty. Accordingly, it could also be used to determine trade-off values between rates and dollars. The licensor can take either a running royalty or an equivalent lump-sum payment, which can then be invested. Again, assume a market interest rate of 10 percent. This is the opportunity cost or rate that is foregone if the licensor elects a running royalty. It is the rate of return on the money if taken as a fixed sum and invested at the current interest rate.

In another example, a 3 percent royalty is offered by the licensor. An up-front payment to buy the 2 percent can be calculated. At 3 percent royalty, the annual royalty is $120,000, and the present-value Amount column calculates to $455,000. The difference between the Amount at 5 percent and the Amount at 3 percent is $303,000. This figure is the dollar equivalent of a 2-point reduction in the royalty rate. It represents the fair amount a licensee could be asked to pay if he or she wished to have royalty reduced by 2 percentage points.

Even though mathematical formulas may be available to calculate reasonably fair royalties, royalty negotiations are often attempts at reaching agreement based on "what is normal for the industry," "rules-of-thumb," or "gut feelings" about reasonableness. These considerations do not take into account the uniqueness of the new technology, especially in regard to its cost, sales potential, or profitability.

Some negotiators believe that concern with detail hampers discussions. Negotiations can be intense, requiring on-the-spot decisions, and executives prefer to have discretion on setting royalties to have room to exercise expert judgment. The negotiations might or might not culminate in a license agreement. In either case, the executives might be required to submit a detailed rationale for their course of action to their employers.

MODEL FOR NEGOTIATING ROYALTY

The model described below provides a tool for explaining royalty rates.[4] It could be useful for those negotiators who require valuation of multiple variables in their determination of acceptable royalties. The model begins with a starting-position royalty: the high and low for the licensor and the licensee, respectively, using information as follows from the licensing executive:

1. An array of considerations or variables judged to impact royalty
2. Reasonable estimates or dollar valuations of the variables impacting royalty
3. Weights indicating the relative worth of the variables impacting royalty.

Item (2) is a quantity consideration, whereas item (3) is a consideration of relative importance; the importance of a dollar spent on the variable. The latter provides for similar quantities of dollars to have varying degrees of importance for the royalty calculation. Illustrative examples using hypothetical values and weights for a selection of variables are also provided, in Appendix B.

The model responds to the executive concerned with observing the effects on royalty of changes in variable valuations or changes in relative worth. It also has value for the executive dependent on "gut-feelings" about reasonableness. The model will generate supporting valuations and calculations of relative worth for each new royalty. The model can be readily programmed into a computer, thereby significantly increasing its accessibility and ease of use.

Three versions of the model are described. Version 1 presents the situation in which an executive wishes to deal with dollar amounts only, translating these into royalty rate equivalents. In Version 2, the executive chooses to work with variable weights only and does not assign dollar values to the variables. In Version 3, dollars and weights are combined.

Opinions vary regarding which variables create differences in royalties. Forty-six variables are identified in one study. In the study the author sought to identify the variables influencing foreign licensing executives in their pricing of technologies. Questionnaires were sent to 353 executives, and 111 returned usable responses. Among the top most influential variables were,

1. Degree of exclusivity of the license
2. Anticipated total money return
3. Type of license (patent or know-how)
4. Term of license contract
5. Term and type of obligations assumed by the licensee under the agreement (i.e., guarantees, immunity from suit, etc.).[5]

Version 1 of the model is initialized in the following manner. Here, the executive decides which variables to use in determining royalty. (These may or may not include selections from the above.) The assumption is made that dollar amounts can be estimated and assigned to each variable, and the objective is to translate these dollar amounts into royalty rate equivalents. Then, to see how changes in dollars cause changes in royalty:

1. Create a Column A, using any number of variables.
2. Assign a dollar value (in any currency) to each of the variables in Column B.
3. Calculate the total of Column B.
4. Create a Column B(2) to list the new values resulting from the first round of negotiations.
5. Create a Column C, with a total equal to the starting royalty and multiply the starting royalty by the ratio of each element of Column B to the total for Column B o arrive at the corresponding element in Column C. (Note: The elements in Column C should equal a given starting royalty rate, e.g., 5 percent).

The stage is now set for negotiating changes in the dollar amounts of Column B. Changes in Column B are recorded in a Column B(2); values not changed are transferred to Column B(2), as is. The new column creates a Column C(2), the total of which yields the new royalty rate after the first round of negotiation. Each round of negotiation creates a new set of values. Columns for each round of negotiations (i.e., Column B(1), B(2), B(3) . . . have corresponding Columns C(1), C(2), C(3), etc. . . .). This process may be summarized in the following manner:

$$bij/\Sigma bij * cij = c(i+1)j$$
$$\Sigma c(i+1)j = \text{the new royalty rate}$$

"i" represents the column and "j" the variable. The columns also represent the round of negotiation. (i = 1) generates a column of values for the variables that initiate the first round of negotiations. There is an array of "j"s for each "i." If no change is made, the original bij value is retained. So, the value for b(1)j/b(2)j is equal to "1." The b(2)j represents a change from b(1)j and the new royalty brought about by the changes are obtained by summing the column of the elements c(i+1)j. Note, the column count starts at B(1). It excludes column A.

Version 2 may be similarly constructed. Instead of dollar values, weights are assigned. Frequently, dollar amounts are not readily available, and the executive might be more comfortable using weights indicative of a factor's relative importance for royalty, in relation to other considerations. Weights may be assigned to each factor, ranging from 0 - 100, from 0 - 10, or any other range with which the executive might wish to work.

Version 3 combines Versions 1 and 2. It assumes that while dollar values may be assigned to each variable, dollars are more important in some variables than in others when determining royalty. For example, an extra dollar in profitability could be viewed more favorably than an extra dollar in sales.

Here again, the system is initiated by values being assigned to the respective variables, as above. However, because both dollars and weights apply, new columns must be created.

1. Create a Column A using any selection of variables.
2. Assign dollar values to each of the variables in Column B.

3. Assign weights to each of the variables in Column C.
4. Multiply each element in Column B by the corresponding element in Column C and enter the product in Column D.
5. Calculate and enter the total for Column D.
6. Create a Column E, giving it a starting royalty rate for a total, and generate a distribution for it using the ratios of the elements in Column D to the total for Column D (i.e., $e_{ij} = d_{ij}/\Sigma d_{ij} \times \Sigma e_{ij}$, where Σe_{ij} is also the starting royalty, for Column $i = 1$).

The six steps outlined above set the stage for negotiations, and the first round, second round, etc., of negotiations produces new columns B(2), B(3), . . . C(2), C(3), . . . D(2), D(3), . . . E(2), E(3), etc. . . , with calculations for each successive royalty rate given by $e(i+1)j$ where,

$$b_{ij} \times c(\Sigma i+1)j = d_{ij}$$
$$d_{ij}/\Sigma d_{ij} \times e_{ij} = e(i+1)j$$
$\Sigma e(i+1)j$ is the new royalty rate.

Appendix B provides an example using hypothetical data.

DIFFERENTIAL IMPACT OF ROYALTY AND FEES ON LICENSEE OUTPUT

Generally, the amount of royalty realized is a function of sales of the product embodying the licensed technology. Sales yield the fruits of the technological innovation, benefiting the licensor, licensee, and the public at large. The greater the quantity consumed, with price constant, the greater the income for the licensing parties. However, price determines to a large degree how much of the licensed product consumers will buy, and price tends to have an inverse relationship with quantity.

By acquiring a license, the licensee secures a certain degree of market power, a right to practice the technology for commercial purposes, either with no competition under an exclusive license or subject to restricted competition under a nonexclusive license. This condition gives the company a certain degree of control over price setting, which it is then able to use to achieve certain profit objectives. In requiring compensation for the license, an additional cost element must be considered in the licensee's determination of the price that would maximize profit and output.

Different forms of compensation have different consequences for all the beneficiaries of licensing. Output and income are affected in particular depending on whether the payment is based on a royalty rate or is a fixed fee. Two economic models are illustrative. One involves nonexclusive licensing of a cost-reducing innovation in a previously competitive industry. The other involves opportunities afforded the licensee by a new product innovation under an exclusive license. Normal profits for the industry are incorporated in the cost function. In both cases, constant marginal and average costs are assumed.

In the competitive industry, any compensation paid to the licensor that is lower than the savings afforded by the innovation will induce producers to take a license, since the innovation promises a profit margin in excess of the normal industry level contained in the economic cost calculation. However, the model exhibits differing effects for fees and for royalty. In the case of fees, only the average cost function increases, while royalties increase both the average and the marginal costs.

Output thus tends to be higher under an equivalent fee regime than under royalty requirements in the license. Because output in a competitive industry is indicative of company size, the lower level of output under the royalty scheme means that a large number of small producers are operating at the optimal industry output. In the case of this industry, the nonexclusive licensees charge the given competitive price of the industry and enjoy profits thereunder as a result of the restricted access of the license.

The situation is different in the case of a licensee who holds an exclusive license and thus can set the price of the product. Having determined the market conditions, the licensor may require payment of a fee and a minimum output at the point of equivalency between marginal cost and marginal revenue, taking all returns in excess of cost. Since the fee does not affect marginal cost, this required output would be compatible with the licensee's output selection in the circumstances. This is the monopolist profit-maximizing output. Any fee paid that is less than the difference between total revenue and total cost will enable the holder of an exclusive license to produce at the profit-maximizing-output level.

While a fee does not increase marginal cost, a royalty does. As a result the exclusive licensee reduces output to the point where marginal cost and marginal revenue are equal. The licensor, however, may again impose an output quota in order to offset the lost income arising from the royalty's effect on production. To conclude, in the case of the exclusive producer with pricing power, the fee yields higher output than a royalty per unit sales.[6] Actual receipts depend on the elasticity of demand.

In reality, most license agreements contain a combination of royalties and fees. The influence of other considerations comes to bear. A preference for royalty on the part of the licensee, besides a sharing of risk by the licensor, ensures that the licensor will take action against infringers to protect the profitability of the venture and a realization of its full commercial potential.

Such opportunities for gain under licenses generally expire with termination of a patent or divulgence to the public of the trade secret on which the license is based. At this time, any companies under the cost obligations of a license face a competitive disadvantage. In this long-term period, entry into a product line controlled by any particular licensor company is opened to entry by others, who are lured by the opportunity for gain.

Notes

[1] Marcus B. Finnegan and Herbert H. Mintz, "Determination of a Reasonable Royalty in License Negotiations," *Licensing Law and Business Report*, Vol. 1, No. 2 (June-July 1978): 13-24.

[2] Ibid., 13-24.

[3] Ibid., 13-24.

[4] Albert E. Muir, "Rationalizing Royalties," *les Nouvelles* (Journal of the Licensing Executives Society) Vol. XXI, No. 2 (1986): 98-103.

[5] Michael D. Rostoker, *A Survey of Corporate Licensing, Patent, Trade Secret, Know-How* (Connecticut: The Franklin Pierce Law Center, 1984), 25.

[6] Christopher H. Hall, "Renting Ideas," *The Journal of Business*, Vol. 64, No. 1 (January 1991): 23-24.

Chapter 8

Developing the License Agreement

When a licensor's marketing of an invention is successful and a company's evaluation produces an interest in licensing, the two parties initiate contract talks. Negotiating skills are the important factor in bringing about a successful license agreement. Generally, the agreement is reduced to writing, has complex legal and business implications, and is an enforceable contract, which means that legal remedies are available in the event of violation by breach or default.

While it is not required by law that the license be in writing, this generally is the case and certain requirements must be satisfied to make the contract enforceable. The agreement usually represent the outcome of careful negotiation. A number of considerations influence the latter as well. In the license agreement, the provisions are carefully organized, and facilitating formats have been developed.

ELEMENTS OF AN ENFORCEABLE CONTRACT

To be a legally enforceable contract, a license agreement must include the following six essential elements.

1. An Offer
2. An Acceptance
3. Mutual Assent by the Parties to the Terms of the Contract
4. Financial Consideration

5. The Object of the Contract Must Be Lawful
6. The Parties Must Be Legally Competent To Enter into a Contract.[1]

An Offer

An offer entails one party stating what he or she is willing to do or provide if the other is willing to commit to a defined undertaking. For example, a prospective licensor might offer a potential licensee an exclusive license if the licensee is willing to undertake the commercialization of an invention.

An Acceptance

The party receiving the offer must accept it. In the case of an offer of an exclusive license for the purpose of commercialization, the licensee must agree to receive an exclusive license. Both parties must accept and be willing to be bound by the terms required by the license.

Mutual Assent by the Parties to the Terms of the Contract

By signing the contract, the parties assent to all its terms. However, the agreement does not protect mutual mistakes. In a mutual mistake, a court would rule that there had not been mutual assent. However, if one of the parties fails to read the agreement before signing or misinterprets its meaning, a court is not likely to release that party from its obligations under the contract.

Financial Consideration

The crux of a typical license agreement is an exchange of commitments between licensee and licensor. The licensee usually promises to pay the licensor a royalty. In return, the licensor grants to the licensee permission to use intellectual property rights belonging to the licensor for the purpose of making profits through commercial sales. The licensor also agrees to prohibit others from making use of the same technology for the same purpose. Restricting use limits competition, thereby providing the licensee with an opportunity for higher-than-usual profits.

The Object of the Contract Must Be Lawful

A contract is illegal and unenforceable if its formation or performance constitutes an unlawful act. The intention of the parties must not be a violation of the law. For example, licenses may cause violations of the antitrust laws, especially of the Sherman Act and the Clayton Act. Other statutes that apply are the Robinson Patman Act and the Federal Trade Commission Act.

A licensor who holds a patent may be in violation of the law if the license imposes a restriction or an obligation on the licensee under the leverage of the patent that the licensor would not otherwise have been able to exert. Such violations can include tying arrangements, package licensing, price restraints, or prohibition of the sale of competing products.

The Parties Must Be Legally Competent to Enter into a Contract

All persons legally of age and not legally disabled may enter into binding contracts. This includes corporations, which are considered legal entities. When the parties sign the agreement, they authenticate the written document. In the case of a corporation, the authority to sign is generally restricted by the firm to certain individuals.

NEGOTIATIONS

Reaching agreement or settlement on the terms of a license is usually achieved through negotiation. The desire on the part of both sides to reach agreement is the cooperative element of negotiation. But negotiations include a competitive element.[2] Each side wants to maximize its own return (reward) at minimum cost to itself. Reward and cost relate to the proceeds and obligations of the license agreement. The successful license is one that pleases both parties, benefiting both in the short term and in the long run.

As a dynamic aspect of behavior, affecting relationships of people and organizations, negotiation and its outcome are significantly determined by human variables. Important human variables are personality and skill. These may be characterized as the internal variables. They determine how one communicates, what one communicates, and what one accepts. Negotiations are also affected by external variables. Important among these are human and organizational influences bearing on the negotiator, negotiation tactics, information, and time.

The internal and external variables affect the atmosphere of the negotiations, which in turn determines the tone of the negotiations, that is, whether the negotiations will be friendly, formal, indifferent, adversarial, or hostile.[3] They also determine the quality of the preparation of the negotiation, resourcefulness, initiatives, reactions, and the outcome.

Personality

Personality is a combination of attitudes and feelings in the individual and how they are allowed to influence behavior. It is reflected in behavior patterns, which are influenced by emotional needs, economic needs, and values. Personality governs to a large extent a person's aspiration levels and reactions to conditions and to the behavior of others. It determines whether the negotiator will tend to be domineering, submissive, or collaborative; vulnerable to the exercise of power by others; and/or willing to use power advantages to achieve objectives. It influences a person's setting of targets and commitment to goals.

Skill

In the context of negotiating a license agreement, skill means competence for the task and ability to understand and deal productively with the variables indicated, within oneself and in others, and with the external variables. Although a license agreement includes technological and legal considerations, the transaction is essentially an economic exchange.

Accordingly, skill involves an ability to research, understand, and communicate the business, market, and economic aspects of the transaction; to recognize technological and legal implications, and to use external sources of help effectively when needed. It involves knowledge and experience in planned, purposeful negotiation with both sides.

Human and Organizational Influences

Human and organizational influences are the organizational, in-group, and/or the personal influences of others that may be available to help the negotiator or that may be brought to bear upon the negotiator to influence behavior. A serious commitment to negotiation by a supportive employer can provide the negotiator with needed resources and flexibility, enabling him or her to make thorough preparation for the task. Alternatively, the employer or client might argue for expectations that are not in accord with the realities of the case, thereby severely compromising the flexibility of the negotiator.

Information

Information is that body of relevant data, methods (including negotiation methods), and knowledge external to the negotiator that can affect the achievement of objectives. It is available in varying degrees to the negotiators on both sides. Both the skill of the negotiator and the accessibility of the information influence the degree of its availability. A skillful negotiator is more likely to be able to identify and process information to advantage. However, information internal to the opponent's organization is generally not as available to the negotiator as it may be to the opponent's negotiator. Effective use of information takes skill; it also builds experience, a determinant of skill.

Time

Generally, negotiations are subject to time constraints. The transaction must be executed before the investment in the new product idea can be made and before any income can be earned. Considerations of efficiency might pressure a licensor negotiator to consummate the deal in order to move on to other projects. The licensee negotiator might be pressured by the company's R&D people who are waiting impatiently to get started. The party for which time is a more pressing issue can be at a disadvantage, particularly if this is known to the opponent. Time pressure can force one side to make concessions not originally intended.

Negotiation Tactics

Negotiations have both short-term and long-range implications. An agreement that leaves one or both parties dissatisfied puts the invention at risk of becoming a failed new product. A sampling of negotiator types includes the demanding negotiator who must win at all costs, the reasonable negotiator, and the collaborative negotiator. Most negotiators will use combinations of these methods, depending on personality and the issues at stake.

The demanding negotiator makes excessive demands and is reluctant to make concessions. The opponent is generally viewed as an adversary and seen as selfish. A combination is a demanding negotiator who is also reasonable. This negotiator begins with realistic demands in the interest of good-faith give-and-take. When two such negotiators are matched, each yields on some of his or her demands in order to realize important objectives. Combining the traits of the collaborative negotiator, the parties make an effort to understand each other's needs, exchanging information in the interest of a mutually rewarding result.

Preparation for Negotiation

Knowledge of the above variables, their combinations, and their use is important in the implementation of strategy and is helpful in enabling recognition of the other party's strategy. It improves the ability to respond effectively to changing conditions and to the tactics and ploys of an opponent. The variables can also be used as guidelines in planning for negotiation.

A plan comprises the objectives to be realized, strategy to be used, and rationale for the desired outcome. Strategy is the means by which the objective of the negotiation is reached; the manner in which variables and tactics are selected. Tactics are the particular uses of the negotiation variables, the more important of which are noted above, to influence the opponent and the outcome.

An important component of preparation is fact-finding and analysis of information. This includes information about the other negotiator as well, such as behavioral predisposition and skill level. The negotiator should also assess the various expectations of his or her employer/client, team members and others with an interest in the outcome, with a view to developing common ground. From this, the negotiator can determine the time constraint necessary for scheduling and implementing the process.

Preparations for negotiation should begin well in advance, at least during the invention evaluation phase. Opportunities are already available for doing so at this early stage. For example, company personnel in charge of licensing usually make their appearance at the time that a new technology is formally selected as a new product/process candidate. An opportunity for the licensor to get to know the legal representatives exists during communications regarding the confidentiality agreement under which the proprietary information is disclosed. The licensor usually meets with the company's R&D personnel during the exchange of technical information.

The review process requires communications between the parties. The negotiator can accumulate information on attitudes, background, general expectations, who the important players will be, decision makers and internal/external pressures. With this background, the negotiator can develop a realistic negotiation plan, based on best guesses regarding the opponent's likely behavior, expectations and tactics.

Additional information pertinent to license negotiation relates to the advantages and disadvantages of the technology. The legal protection available to the invention should be assessed, including the relevance of significant patents. Information regarding technical and economic conditions also needs to be assembled and analyzed, as does the compatibility of the new product/process with existing product line(s), its fit within corporate know-how, its advantages over competitors, its potential impact on business earnings and forecasts, and its likely return on investment. Since a license is a long-term relationship, the fact-finding is not complete without a record of the new partner's licensing history and the quality and duration of other license relationships.

Negotiation Process

The best negotiations generally begin on the most positive note. It is particularly helpful to start with issues on which the parties are in general agreement. Progress can then advance, beginning with the least troublesome issues and allowing a development of a relationship able to overcome the more difficult issues. Note, however, that the least troublesome issues are not necessarily the least important ones.

Another useful place to start is with an agreement form, which usually has legal and administrative provisions of general acceptance. More case-specific terms can then be proposed. The initiating party under this format enjoys the advantage of establishing the reference point for the negotiations. If prenegotiation preparation has been thorough, the negotiator will be prepared to argue substantially for the terms proposed.

An important aspect of progress is flexibility. This is facilitated if the negotiator has defined boundaries within which he or she will work. If possible, the negotiator should avoid being categorically negative in responding to a provision. Rather, the negotiator should provide explanation of why the provision is unacceptable and should suggest corrective action that might be helpful. When the opponent is rigidly negative, the negotiator should encourage discussion toward a resolution, possibly offering inducements if necessary.

Concessions should be gradual and counterproposals well timed. Immediate, excessive concessions can be unproductive, particularly if the opponent is unprepared or if the concessions do not reflect the opponent's priorities. Counterproposals should show a responsiveness to the opponent's proposals, and conveyed as an inducement to progress. The negotiator should not rule out the possibility of terminating negotiations if the other party is not being responsive to important objectives. The sobering effect of this knowledge can induce productive reformulations of positions.

THE LICENSE AGREEMENT

There is no law that states that a license must be expressed in the form of a legally enforceable contract, but this is generally the case. Careful documentation should be kept of all matters addressed, resolved, and agreed upon, so that both sides understand the terms of the license. This is also necessary in order to avoid misunderstanding and controversy later on.

The parties make commitments in a license agreement that endue for years and that have far-reaching implications. Breach or default of the agreement can have disastrous consequences. Not surprisingly, therefore, the parties rarely contest the necessity of having a written and signed document enforceable by law to represent the understanding. Negotiations are conducted and a license agreement is formulated within this context.

License agreements may be based on patents, know-how, and/or combinations thereof. Patent licenses relate to technology that is in the public domain, under the protection of patent law, whereas the technology of know-how licenses is outside the public domain and is protected by trade secret law. Both patents and know-how can be protected under exclusive or nonexclusive licenses.

Under patent rights, the patentee holds the right to exclude others from practicing the invention. This right is waived for the licensee in a patent license. On the other hand, the technology of a know-how license is not in the public domain, and an obligation of confidentiality in the agreement serves to preserve the advantage enjoyed by the practicing parties.

Often a license agreement is made while a patent application on the new technology is pending. Since a patent does not yet exist, such technology must rely on trade secret law for its protection. Also, these licenses typically provide for the possible outcomes of the application: granting of a patent or rejection of the application. This contingency is often reflected in two royalty scales: one royalty rate for a know-how case and a higher royalty rate for a patent.

The following format is relevant for licenses, irrespective of the intellectual property designation of the technology. As the sample document in Appendix C-3 indicates, the provisions of the license, or the subject matter agreed upon, is arranged under numbered and titled Articles to facilitate reading and reference. Typical inclusions are the following:

1. Introduction
2. Definitions
3. Grant of License
4. Due Diligence
5. Royalty
6. Reports and Payments
7. Termination
8. Unlicensed Activity and Infringements
9. Assignment
10. Hold Harmless
11. Communications
12. Miscellaneous
13. Signatures

Introduction

The agreement opens with a date, identification of the parties, and recitals. If the effective date is not specified, the initiating date for the contract is generally the signatory date that the last of the parties sign the agreement. Full names and addresses are required. For corporations, this includes the state of incorporation. The recitals, or "Whereas" statements, are used to convey background information relating to the transaction but are not essential.

Definitions

Terms essential to the agreement are presented and defined in the Definitions Article. These are capitalized as they are defined and are also capitalized when they are used in the agreement. This section usually includes a specification of who the licensee is and the respective affiliates to be granted access to the technology; a definition of the technology licensed; the product it produces; and the basis of royalty calculations, such as net sales. Less important terms requiring explanation are generally defined in the agreement as they occur.

It is particularly important that the definition of the technology and its application in the license leaves no ambiguity as to the scope or restrictions of the grant. Often, patents that have not been granted, patent applications that have not been filed, and/or relevant technology to be developed are referenced. For these cases, it is customary to attach a schedule that can be amended periodically as the updates become available.

The Grant

This Article articulates the grant of the agreement. The licensor grants to the licensee the right to "make, have made, use, sell, or lease" products embodying the technology. In this grant, the licensor waives the right to exclude the licensee from practicing the technology for commercial purposes. The Article specifies whether the license is to be exclusive or nonexclusive and whether the license is based on a patent or know-how. The license is also given its time frame and the territory over which rights apply. For exclusive licenses, provisions for sublicensing are indicated.

Royalty

The payment or consideration for the license and its form is presented here. Generally, the royalty is tied proportionately to the use of the technology, usually as an amount equal to some percentage of sales. If the license includes an exclusive and a nonexclusive phase, separate royalties are specified. This is also the case if the license converts from a patent to a know-how basis. In addition, a separate rate is applied to income earned by the licensee from sublicensees.

Due Diligence

The licensor is entitled to an assurance from the licensee that the technology will in fact be practiced for commercial purposes, assuming that this is the intention of the license. It happens occasionally that a company will take a license in order to foreclose the market entry of a competitor product.

By stipulating a minimum standard of performance, the licensor is able to salvage the technology in a termination due to breach of contract if the stipulated due diligence is not exercised. Such a stipulation might include specification of milestones in the progression of the licensed product to a commercially ready stage, funding levels to ensure such progress, a deadline date for first commercial sales, and/or a minimum annual sales amount.

Reports and Payments

Indicated here is the requirement that records pertinent to a royalty accounting be kept for a certain time period at a specified location. The location is usually the headquarters of the licensee. The records are subject to inspection by the licensor's personal or independent auditor. The Article usually requires that reports based on these records be prepared and submitted to the licensor periodically, and the items to be included in such reports are indicated. If sublicensing is involved, copies of reports to the licensee from its sublicensees are generally required.

Normally a report is due shortly after the last day of the royalty period, at which time royalty is due and payable. If a foreign currency is involved, it is customary to specify that the official exchange rate be used for conversion to U.S. dollars. If no official exchange rate exists, then the bank buying rate at some major city is used.

Termination

Termination occurs when the license expires unless there is a provision for renewal. This may be the case if the license period is shorter than the life of the licensed patent. Bankruptcy is also a basis for termination.

The injured party in a breach or default may terminate the agreement. The licensee, for example, might terminate an agreement if the technology does not live up to its claims. The licensor might terminate it if there is a breach of due diligence or a default in royalty payments. Generally, however, a grace period is allowed for the offending party to cure the breach or default.

Obligations associated with the license do not instantaneously expire upon termination. Licensed products still in process of manufacture and inventory must be sold. A time frame for such sale is normally allowed, and royalty continues as payable under the license terms.

Unlicensed Activity and Infringements

Generally, the licensor is the party most concerned about infringement. Not only is the patent the basis of income for both parties, but it can be argued that the licensor has an implied obligation to protect the licensee if the licensed patent infringes another patentee's rights. The infringement provisions recite actions for the parties and their respective participation in law suits, expenses, and recovery of damages.

With respect to actions, the parties share a mutual obligation to inform each other promptly of any infringement. They have an obligation to cooperate with each other in the defense or enforcement of the licensed rights. The course of action to be taken is the prerogative of the licensor. However, if the licensor fails to act or is unsuccessful, then the licensee may take action with the permission of the licensor. The agreement of both parties is required in a final voluntary disposition of the case.

Costs are borne by the party commencing or defending the infringement. The licensee may withhold royalties as reimbursement for its expenses, and any recovery of damages in excess of unreimbursed expenses and legal fees is shared proportionately with the licensor. If the settlement entails payment of royalties or damages by the licensee, a commensurate reduction is made in the royalty obligations of the licensee to the licensor.

Assignment

A license agreement is necessarily affected if the licensee sells commercial rights to the license or merges with another corporation. In the case of a sale, the licensor may miss the opportunity to partake in a significant payment. Such a transfer, or a merger may have the undesirable effect of delivering the technology to a competitor wishing to remove the product from the market. The license agreement is also affected if the licensee is a person who dies, thereby releasing rights to heirs or legatees. Heirs or legatees might not have the business interest or inclination to continue the diligent practice of the technology.

The above events illustrate the potential risk of a transfer. Therefore, it is important for the licensor to retain the option to carefully evaluate the attending costs and benefits of a contemplated assignment, the discretion to approve or disapprove of a prospective transfer, and the right to terminate the license in the event of a disapproval. The license should also provide an option for the licensor to participate in proceeds received by the licensee in exchange for the assignment.

Hold Harmless

The licensee bears the responsibility to produce and market a safe, reliable product under the license. In the event of injury or economic loss due to a faulty product, consumers can bring charges against the licensee as producer of the licensed product. Nonetheless, licenses frequently provide language expressly relieving the licensor from any obligations to final buyers. As a further safeguard, the licensee may be required to name the licensor as the insured party. However, the licensor has an obligation to deliver to the licensee a product that is compatible with the specification of the patent and should be prepared to warrant its technical correctness and effect.

Communications

Royalty payments; notices of infringement, breach, default, or termination; and other such important mail can be at risk of being lost in transit. This risk and the uncertainty regarding the effectiveness of a notice is reduced if the license agreement expressly requires a safe method of transmittal and indicates the addresses to which mail should be sent. A clause stating the time when a notice takes effect is also useful.

Miscellaneous

This is a catch-all Article for items. Items here can alternatively be treated as separate articles, for example, applicable law. Laws differ not only among countries but also among states in the United States. To avoid problems later on, it is customary to identify in the license the country or state whose law is to apply.

Signatories

The legitimacy of the license agreement is dependent on the authority of the signatories to bind the contracting parties. While it may be self-evident in the case of individuals who know each other that the person signing is also the person to be bound, such authority in corporations is restricted to a select few individuals, usually officers of the company. However, most of the time, such individuals become known during the progress of the transaction. Notarized signatures and/or corporate seals are means sometimes employed to ensure that the right party is signing.

Notes

[1] John R. Goodwin, *Business Law Principles, Documents and Cases* (Illinois: Richard D. Irwin, Inc., 1976), 157-175.

[2] Alan N. Schoonmaker, *Negotiate to Win Gaining the Psychological Edge* (New Jersey: Prentice Hall, 1989), 5.

[3] Ibid., 52-63.

Chapter 9

Applying for a Patent

Patents are nation-specific instruments. An inventor must file a patent application in each individual country in which a patent is desired, in accordance with that country's laws, and must pay the respective filing costs, translation expenses, legal charges, and other costs. However, the applicant may pursue patenting rights in multiple countries by filing one application if the filing is done in accordance with applicable international treaties or conventions.

Before filing a patent application, the applicant can determine the likelihood that the patent will be granted by conducting a patent search. Initially, the applicant has a choice of filing either a provisional or a regular application. In both instances fees must be paid, and certain documents must be filed in the United States Patent and Trademark Office (PTO). At this time the applicant may also wish to consider whether foreign rights ought to be pursued, and the available options.

THE PATENT SEARCH

To constitute patentable subject matter, the invention must not already be in existence; must not be evident in issued patents, published materials, or other public documents or disclosures; or must not be readily deducible from such disclosures. However, this prior art does not include prior inventions in foreign countries that are not the subject of patents and patent applications. A patent search involves the examination of this prior art to determine patentability.

A valuable resource available to the prospective applicant is the Search Room at the United States Patent and Trademark Office (PTO), which is open to the public. The Search Room contains copies of patents since 1836. It also has many thousands of volumes of technical books, scientific journals, and periodicals, along with publications of foreign patent organizations and millions of foreign patents.

Copies of patents are also kept at Patent Depository Libraries which are located throughout the country. Besides copies of patents, these libraries have technically trained people to help searchers make full use of the available library resources. Patents are filed under a classification system that includes more than 400 classes and 115,000 subclasses of patents. An on-line computer data base known as the "Classification and Search Support Information System (CASSIS) identifies the classifications.

THE PATENT ATTORNEY

In pursuing a patent, it is always advisable to work with a competent patent attorney or patent agent, who has a background in the field of the invention. A common source document in locating an attorney is the yellow pages in the telephone directory. Also, the United States Patent and Trademark Office keeps a public record of all patent attorneys and agents admitted to practice before the Patent Office. *A Directory of Registered Patents and Agents Arranged by State and Counties* is sold by the Superintendent of Documents.[1]

In order to empower the attorney to act in his or her behalf, the inventor must sign a document called a Power of Attorney. The inventor must provide the attorney with as complete a description of the invention as possible, including drawings. This will be the reference material for the patent search and application. The inventor is also under an obligation to disclose information of which the inventor is aware that may be material to the examination of the patent application by the patent examiner.

THE PATENT APPLICATION

The most basic initial protection available to the invention is the disclosure document. While this filing does not constitute an application, it may be accepted as evidence of the date of invention, and is maintained by the PTO for two years. The application choices available are, the provisional application and the regular application. Timing in both instances has significant implications for establishing priority as well as longevity for the respective patent(s). Prosecution time is an additional factor, since the 20-year life of the patent is measured from the date of filing of the patent application. Delays in PTO Actions and patent attorney responses also have the effect of shortening the life of the patent. To realize the full benefit of the patent term, action in these areas must be expeditious and without delay.

The Provisional Application

Before filing a regular patent application, an applicant has the option to file a provisional application. Under this procedure, an applicant can quickly file a preprint manuscript or student thesis without incurring the time or expense of a regular application. The respective regular application however, must be filed one year after the filing date of the provisional. At that time, one year after the provisional filing date, the life of the provisional application expires. This application is not examined for patentability.

By first filing a provisional application, the applicant effectively establishes a legal claim to the technology without initiating the life of the patent. Public disclosures of the information filed do not result in loss of patenting rights in WTO member countries, unless it is allowed to go abandoned at the end of its term. During the period of the provisional, updates are allowed, and all updates benefit from the priority date established. The rules also provide for the conversion of a regular application to provisional status.

The earlier the filing of the provisional application, the smaller the volume of prior art that can be used against the regular application and the more likely it is that the applicant will have senior party status in the event of an interference. Also, the chances of broad claims foreclosing competitor claims is enhanced the greater the number of specific embodiments incorporated in the provisional filing.

Since this application does not undergo examination, there is nothing to prevent the applicant from incorporating multiple inventions in one provisional, thereby achieving priority for all under one application. Later when the corresponding regular applications are filed, separate filings will need to be made. In filing an early provisional, the advantages that accrue to one invention are automatically extended to all those incorporated in the application, for the same filing fee.[2]

Certain conditions must be satisfied for the benefits of the provisional application to be realized. The cover sheet transmitting the invention is important. Specifically, it must identify the application as a "provisional application," otherwise it will not be treated as such. The cover sheet must contain the names of the inventors, the title of the invention, the name, registration number, docket number, and address of the patent attorney (if applicable). While the filing does not require a recitation of claims, it must be sufficient in detail to enable reproduction of the invention by one skilled in the art. It must explain the best mode of practicing the invention, including any drawings that may be required.

The Regular Application

The Patent Act of 1952 establishes a structure for the application. Under the Act, the application must contain a specification of the invention and a drawing. The sections are comprised of the Title, Abstract, Background of the Invention, Summary of the Invention, Detailed Description of the Invention, and the Claims. A declaration or oath of inventorship by the inventor completes the formal requirements.

In the PTO, the application is assigned to one of a group of patent examiners, who are grouped by area of technology, and is taken up by an examiner in accordance with predetermined Patent Office procedures. Strict rules are observed in the PTO, and any deviation from established procedure must be justified to the Commissioner of Patents.

The examiner then proceeds to determine if legal requirements have been satisfied and evaluates the invention relative to the criteria for patentability: novelty, utility, and nonobviousness. Regarding nonobviousness, the examiner looks at the prior art in domestic patents, foreign patenting activity, and the literature, then renders an Office Action. If the application fails to satisfy the patentability criteria, the application is rejected. There is a first and a second Office Action. The second action can be made final and can be followed by an appeal.

Each application is entitled to two Office Actions. The first Office Action is a written explanation from the examiner to the applicant, with references to the prior art and citations of references as to the reasons for the rejection of some or all of the claims of the application. The relevant law is also indicated.

The applicant may respond to the Office Action with an amendment in which an attempt is made to overcome the objections of the examiner. It is also possible to arrange an interview with the examiner, but this does not remove the necessity of the official response, since only the written record applies. If the applicant is successful, the result is a patent issue. If not successful, the result is the second Office Action. Like the first, the second Office Action is a detailed explanation of the examiner's objections.

The applicant may respond to the second Office Action by filing a Notice of Appeal. If the examiner is unyielding, an Appeal Brief may be filed with the Board of Patent Appeals and Interferences in the Patent Office. This may then entail an oral hearing. If the Board rules against the applicant, the persistent applicant may then file a civil action against the Commissioner of Patents and Trademarks or may appeal to the Court of Appeals for the Federal Circuit. Ultimate recourse is the Supreme Court.

If the examiner determines that two or more inventions are revealed in the application, the applicant is issued a restriction order. This advises the applicant to choose one of the inventions for coverage by the application. If one or more additional applications are subsequently filed as a result of this right to "divide" the application, each new application is called a Divisional Application, entitled to a separate patent. All get the benefit of the original filing date if applications are filed during the pendency period.

Other types of applications that emanate from the prosecution process are the Continuation Application and the Continuation-In-Part ("c.i.p.") Application. The former is a substantially unchanged refiling of the original application. The c.i.p. application is used if additional information is provided as a result of testing, use, or experimental work to either expand or broaden the claims of the original application.

Interference

Sometimes the PTO receives separate applications for essentially the same material from different applicants. This gives rise to an interference. The PTO "declares" an interference in order to determine the true inventor. Also, an interference can be requested by an applicant if the PTO inadvertently issues a similar patent to another inventor, provided the date of the still-pending application is no more than one year prior to the issue date of the patent. It is also possible for the owner of the pending application to incorporate claims of the issued patent in his or her application.[3]

Contenders seeking to establish themselves as the first inventor may submit as evidence inventive activity in any WTO (World Trade Organization) member country. With respect to activity in foreign countries, admissible dates are those occurring one year and after January 1, 1996. This is the anniversary date of the implementation of the WTO agreement.[4] However, the contenders cannot rely on such evidence if the invention relied upon was subsequently abandoned, suppressed, or concealed.

If both applications are allowed before detection of the similarities, the patentee may file an application for a reissue of the patent, copying the claims of the other, with a request that an interference be declared. In this event and the above where an applicant requests an interference, the burden of proof is on the last to file, called the "junior party." The first to file, or the "senior party," is able to rely on his or her filing date as the date of the invention.

The interference procedure is conducted by the applicant attorneys. It involves a preliminary statement, a motion period, and a testimony period. In the first instance, the respective positions of the parties are presented. Next come challenges to the rights of each. Written records are crucial in this process. During the testimony phase, the attorneys examine witnesses under oath in the presence of a notary public or other individual qualified to take sworn testimony.

If the applicant is not satisfied with the ruling of the PTO, an appeal can be filed with the Board of Patent Appeals and Interferences. Transcripts of the proceedings are then submitted to the Board of Appeals and Interferences, which decides who is entitled to the patent. The case may also be appealed before the Court of Appeals of the Federal Circuit if this latter action is unsuccessful, or civil action can be taken against the successful applicant. Alternatively, the parties may decide on a settlement in which the loser obtains a license from the winner. Upon resolution by the Board or settlement, the case is returned to the primary examiner who then issues a Notice of Allowance.[5]

Deposit Requirements Concerning Biological Material

For some inventions, the applicant must go beyond a comprehensive technical description. This is generally the case in regards to microbiological and genetic engineering technologies. In addition to a full and enabling disclosure required by 35 USC 112, the law requires a deposit of the subject specimen in an approved repository and a notification of this location, including the access number for the deposit, in the patent specification. The provision is made in order to ensure disclosure of the "best mode" and to enable one of normal skill to practice the invention without undue experimentation.

The deposit requirement applies to pertinent microbial culture as well as genetically engineered microbes, including plasmids and cloned genes. If the invention involves a microbe as the host for a foreign DNA, the microbial host is deposited. While a specimen such as a plasmid DNA might not require a concurrent deposit of its host, there is a question as to whether the applicant is providing the best mode by failing to make this deposit.

The services of a reputable repository should be obtained for the deposit. Here, the material should be maintained in its viable and productive mode, in strict secrecy, during the patenting process. Should the application be abandoned, the inventor can then reclaim the specimen for practice of the technology under Trade Secret Law. Suitable repositories in the United States are the Agricultural Research Culture Collection in Peoria, Illinois, and the American Type Culture Collection in Rockville, Maryland.

The United States is also a member of the Budapest Treaty on the International Recognition of the Deposit of Microorganisms for the Purposes of Patent Procedure. Under this treaty, a single deposit in an approved repository satisfies the patent application disclosure requirements of all the member countries.[6] However, in the event of foreign applications requiring public disclosure of the application prior to patent issuance, the deposit must become publicly available as well. This necessarily causes a loss of the inventor's option to use the invention as a trade secret should the application be subsequently abandoned or denied.

DOCUMENTS AND FEES

Various documents must be filed with the PTO during the application process. The basic ones relate to the granting of legal authority to act in one's behalf; the size of the applying entity, since different fee structures apply; and the declaration of inventorship. Numerous fees must be paid. These include basic charges, penalties, and fees for exceptions to the rule that the applicant may require.

Documents

The Power of Attorney must be filed in the Patent Office, and all communications are then conducted with the so-appointed attorney or agent. Nonetheless, the inventor is free to contact the Patent Office regarding the status of the application.

The law also requires a filing of an official declaration of inventorship. In this document, the inventor is represented as the first to discover the subject matter of the application. As an alternative, the allegation of inventorship may be delivered under an oath sworn before a Notary Public. However, notarization is not required for a declaration.

The Patent Office distinguishes between "small entities" and other applicants and provides a reduced fee schedule for the small entities. Generally, an independent inventor can file as a small entity and so be entitled to lower fees. Each person or organization having rights to the invention and claiming small entity status must file a separate statement to qualify for the reduced fee. This includes assignees.

The inventor may transfer some or all of his or her interest in a patent application to another person or organization. Frequently this is the case for the employee inventor. For purposes of the PTO, such transfer occurs under an assignment, which is a written document executed before a Notary Public.

The above transactions may be noted on a preprinted form available from the Patent Office and comprising the Patent Application Transmittal Letter, which serves to transmit the patent application, the documents indicated, and applicable fees.

Payments Required by the PTO

Patent Office fees and charges fall under several categories. The categories are further broken down into subcategories indicating the many conditions giving rise to payment obligations. Several differential schedules apply for small entity and other than small entity status applicants.

Filing Fees — The utility application contains a basic fee component and an additional fee component. Twenty claims, including three independent claims, are allowed under the basic fee. Additional fees must be paid for claims exceeding 20 and for each independent claim above three. Multiple dependent claims also require the additional fee, so do claim presentations made after the original filings.

Patent Application Processing Fees — Receipt of the application by the Patent Office and satisfaction of the deposit requirements by the applicant initiates the prosecution of the application. However, failure to correct defects may entail additional charges. The applicant is then notified of the serial number and filing date of the accepted application and the examination by a patent examiner proceeds.

In due course, and if some or all of the claims are rejected, the applicant is notified in an "Office Action," stating the reasons for the rejection(s). Two such Office Actions are allowed. In each case, the applicant has the opportunity to respond to the examiner's decision. A time frame for the response is set, which can be extended upon payment of an extension fee. This fee is higher for longer extension periods.

The second Office Action also constitutes the final rejection. However, the applicant may then file a Notice of Appeal to the Board of Patent Appeals and Interference upon payment of a fee. Yet other fees are charged for a brief in support of the appeal and a request for an oral hearing.

International Fees — International filing is facilitated by use of international conventions that countries have adopted. In particular, under the Patent Cooperation Treaty (PCT), applicants may file one international application with the PTO, thereby deferring applications in the individual designated countries. At the later stages, various other charges also apply, such as transmittal, search, and translation fees.

Patent Issue Fees — If the application is allowed, the applicant will receive a Notice of Allowance from the Patent Office, and a Patent Issue Fee will become due. Small entities pay half the amount required of other than small entities. If the payment is not made, the result is a nonissue or an abandonment of the patent.

Postissuance Fees — Following issuance of the patent, errors might need to be corrected. A request for re-examination or a statutory disclaimer may also be filed. In order to keep the patent active for its duration, a maintenance fee must be paid at 3 1/2 years, 7 1/2 years, and 11 1/2 years after the patent grant. The patent becomes abandoned if payment is not made within a six month grace period.[7]

FOREIGN PATENT APPLICATIONS

According to U.S. law, it is necessary to obtain approval from the United States Patent and Trademark Office before filing an application in a foreign country for an invention made in the United States. This filing can be made either before or after an application has been filed in the United States. If the foreign filing is made after the U.S. filing, the filing of the U.S. application constitutes the request for approval. Typically, the grounds for denial reflect national security concerns.

However, no approval is required if the foreign application is made six months after the U.S. application, provided there has not been a denial.

The United States law is different from laws in other countries in a number of important respects. First, the first to invent is the inventor, whereas in other countries the first to file is usually taken to be the inventor. Moreover, the United States allows a one year grace period within which to file an application following an enabling public disclosure of the invention. In most other countries, filing rights are foreclosed by such disclosure.

In the United States, the patent application is kept in confidence in the Patent Office until the patent is issued. If the patent application is subsequently abandoned, the applicant is still able to practice or license the technology as a trade secret, provided there has been no enabling public disclosure. The Patent Cooperation Treaty requires publication of the patent application 18 months after filing in a member state.

International treaties and conventions are designed either to protect rights, to facilitate the applications process, or to reduce costs. Most frequently applicable for applicants in the industrially advanced countries are the Paris Convention for the Protection of Industrial Property (the International Convention), the Patent Cooperation Treaty (PCT), and the European Patent Convention (EPC). This latter convention, the EPC, may be superceded by a European Community Patent Convention (CPC), which must be ratified by all member states before implementation. The EPC results in a bundle of individual country patents, whereas the CPC will result in a true European Community wide patent. Other important conventions are the Pan American Conventions and the African Regional Industrial Property Organization. The United States is a member of the Paris Convention and the Patent Cooperation Treaty.

The Paris Convention

The Paris Convention for the Protection of Industrial Property affords a priority date for applications. That is, once an application has been filed in a member county, and for a period of 12 months thereafter, filing in any of the other member countries is not foreclosed by a public disclosure of the invention or other such act that might otherwise have invalidated application rights.

Subsequent applications in member nations adopt this filing date as the priority filing date. However, member states preserve their sovereign rights as to the determination of patentability, patent issuance and adjudication. Patent rights enjoyed by the citizens of a member nation granting a patent are extended to citizens of other member nations acquiring patent rights in that country.

The Patent Cooperation Treaty

This treaty allows applicants in different countries to file one patent application in a member state. At this filing, the applicant must name all the PCT countries in which application is sought. The application may also include the European Patent Organization (EPO) as a "selected county." A PCT filing requires absolute novelty of the invention that is, it must not have previously been publicly disclosed either orally or in a publication, unless such disclosure is preceded by a filing in a Paris Convention country.

The PCT application processing is divided into two phases. Chapter I and Chapter II. In the first phase, an International Search Report is produced. Chapter II results in a Preliminary International Examination Report, rendering a patentability opinion. The patentability criteria are, novelty, non-obviousness and industrial applicability. However, this opinion is not binding on the individual member nations. The applicant may also separately request a Preliminary International Examination, indicating the countries where it will be used.

Eighteen months following the date of the PCT filing the application is published, and enters the public domain. Twenty months following the initial PCT filing date, copies of the application, translated where necessary, must be sent to countries the applicant has designated for coverage. When the application is received by these nations, it is subjected to an additional patentability review and an examination.

If the applicant has requested a Preliminary International Examination prior to the end of the nineteenth month following the priority date, the deadline for submission of the international application to the individual designated states is extended from twenty to thirty months. The patent is issued, if the application is allowed, in accordance with the law of the respective country.

The European Patent Convention

The European Patent Convention (EPC) is a multi-national agreement between the independent nations of the EPO. Under the EPC, an applicant for a patent may file a single patent application covering member countries designated by the applicant at time of application. The filing occurs at The Hague or in Munich, and is made in one of the three official languages: English, German, or French. Payment of annuities is required during the pendency of the application, and national fees also become due when the patent issues. The latter patent maintenance fees are payable in the individual designated countries after the patent issues.

The application is then examined by the European Patent Office to ensure compliance with filing requirements, and a prior art search is made. As with the PCT, the rule of absolute novelty applies. On completion of the search, the European Patent Office issues a Search Report. The search results and the patent application are subsequently published, 18 months following the priority date.[8] Eventually, if allowed, a group of national patents is issued for the designated countries. Any member state can then require translation of the patent into its official language. The life of the patent is twenty years from the date of filing.

Notes

[1] U.S. Department of Commerce, *General Information Concerning Patents* (U.S. Department of Commerce, 1986), 8-10.

[2] Peter Dilworth, "Some Suggestions for Maximizing the Benefits of the Provisional Application," *Journal of the Patent and Trademark Office Society*, Vol. 78, No. 4 (April 1996): 233-238.

[3] U.S. Department of Commerce, *General Information*, 8-24.

[4] Charles E. Van Horn, "Effects of GATT and NAFTA on PTO Practice," *Journal of the Patent and Trademark Office Society*, Vol. 77, No. 3 (March 1995): 108-120.

[5] John T. Maynard, *Understanding Chemical Patents: A guide for the inventor* (Washington, D.C.: American Chemical Society, 1978), 67-74.

[6] Roman Saliwanchik, *Legal Protection for Microbiological and Genetic Engineering Inventions* (Massachusetts: Addison-Wesley Publishing Co., 1982), 45-68.

[7] U.S. Department of Commerce, *General Information*, 33-38.

[8] Alan J. Jacobs, ed., updated by Elizabeth Hanellin, European Community Patent Convention, *Patents Throughout the World* (New York: Clark, Boardman, Callaghan, 1994), E-21 to E-31.

PART III

Oversight

Chapter 10

Technology Transfer Management

For the present purpose, technology transfer management is the means by which an institutionally-owned inventions portfolio is managed with regard to marketing, patenting, licensing, and administration. The operating unit charged with this function is generally part of a larger multipurpose entity, such as a government agency, university, or nonprofit corporation. An example of one such larger entity is The Research Foundation of State University of New York (Foundation).[1]

At the Foundation, technology transfer responsibilities are defined under a policy instrument, and the manner in which they are managed is subject to organizational guidelines. In its effort to decentralize services, the Foundation has created a number of technology transfer units for managing the inventions of the multiple campuses of State University of New York (SUNY). However, due to the costliness of such functions other campuses continue to share the services of one office at the Foundation's headquarters in Albany.

BASIS OF THE OFFICE

Technology transfer at SUNY and at the Foundation is governed by the Inventions and Patent Policy of the State University of New York. In its first paragraph, the Patent Policy states:

> . . . While carrying out its research mission, State University further recognizes that inventions of value to the public will be made by persons working in its facilities. It is the policy of State University to encourage such inventors and inventions and to take steps to aid the inventor and ensure that the public receives the benefit. Appropriate steps include securing research support, identifying inventions, securing appropriate patents, marketing inventions through licensing and other arrangements, and managing royalties and other invention-related income. These activities are undertaken in a spirit of cooperation with governmental agencies and private industry as part of State University's contribution to the economic well-being of the State of New York and of the nation. . . .[2]

The Foundation's Technology Transfer Office, which is comprised of a central office and decentralized campus offices, was created for this purpose (TTO, or Office). Its particular objective is ". . . to encourage such inventors and inventions and to take steps to aid the inventor so that the public receive the benefit." The Patent Policy defines the Office's responsibilities as ". . . securing research support, identifying inventions, securing appropriate patents, marketing inventions through licensing and other arrangements, and managing royalty and other invention-related income." Paragraph 3(e) goes on to state:

> . . . Nothing in the policy herein stated shall prevent the acceptance of research grants from, or the conduct of research for, agencies of the United States, either directly or through the Research Foundation, upon terms and conditions upon applicable provisions of Federal law or regulations which require a different disposition of inventions or patent rights, . . .[3]

This language extends the responsibilities of the Office to include the management of compliance with sponsor regulations. Outstanding among these is the Federal Government, which provides most of the research support giving rise to invention within SUNY.

CORPORATE MANAGEMENT GUIDELINES

As an operating unit of a larger corporation, the TTO is guided by the Foundation's mission and goals. These are classified under Major Functions:

1. Corporate Function
2. Compliance Function
3. Operational Function

Each Function has a Major Objective, and under each Objective are Core Processes and specific Goals. Activities of the Office will be discussed in accordance with this scheme.

The Major Objective of the Corporate Function relates to those activities that result in providing quality services. It states that the organization (through its operating units) will use

> Continuous quality management principles, including a focus on the customers, actions based on facts, employee participation, continuous communication, and process improvement.[4]

The Major Objective under the Compliance Function is fulfillment of Patent Policy and sponsor policy mandates. The Operational Function's Major Objective is to carry out the specific tasks required by the Patent Policy. Customer Satisfaction is a goal in the execution of all three Functions and it is addressed specifically below.

CUSTOMER SATISFACTION

The Corporate Function requires the Office to identify its customers and to execute its responsibilities in a manner that results in customer satisfaction. A customer may be seen as one who provides value as a response to some inducement or incentive. As will become evident below, the TTO must view a number of parties as customers. These include SUNY inventors, business, the Foundation and the public.

By managing an invention from the time it is disclosed through its life under a license, the TTO accomplishes SUNY's commitment to participate in economic development. As stated in the Patent Policy, the activities of the Office ". . . are undertaken . . . as part of the SUNY's contribution to the economic well-being of the State of New York and of the nation." Therefore, the collaboration between the Foundation and the business community benefits the inventor, the business, and the general public.

The benefit to the public of this collaboration of business and the Foundation is realized in new products, processes, and technological progress. In payment thereof, business receives income from sales to the public. SUNY serves the public as specified in the Patent Policy and receives payment in the form of continuing funding for research from tax revenues, that is, from the public.

The Inventors

Inventors provide value which is contained in their inventions when they disclose them to the Office. In this they exhibit a response to a number of incentives and inducements. While disclosure is required as a condition of employment, the Patent Policy calls for inducements to ensure such disclosure and the incentives are significant motivating factors.

In exchange for disclosure, the offerings include the Office's marketing, licensing and patenting services, which are cost free to the inventor, and a participation in proceeds from licensing. Under the Patent Policy an inventor receives 40 percent of the gross proceeds from the licensee(s) of his or her invention. The possibility of being named inventor on a patent is another major inducement to disclosure, as is the promise of achieving name recognition through the licensing process for technological contributions in products on the market.

Additionally, an opportunity to become better known in the business sector is also provided through TTO marketing efforts. These efforts enhance the inventor's opportunities for industrial research collaborations and support. Furthermore, the Office must provide the inventor with reports on actions detailed under the Operational Function, regarding his or her particular invention.

As customers, inventors can expect active support from the Foundation, since ownership rights of the Foundation depend upon the TTO's continuing attention to the marketing, licensing, and patenting of inventions by the Office.

Business

Through its marketing efforts, the TTO provides businesses with broad exposure to a host of new inventions. By so doing, awareness of new product or process opportunities is not restricted to a privileged few. Broad exposure also increases the potential for business collaborations with the University and its research faculty.

However, in its marketing program the Office must induce business to notice and accept particular SUNY inventions for commercialization. The inventions are among hundreds of thousands from different sources which business must evaluate and which compete for development into new products and processes. These submitters, competitors with TTO, desire to realize the returns on the investments made in the development of their inventions, which come in royalty payments when successful commercialization is realized.

The TTO service to business as customer therefore is the access provided in its marketing program to new product/process opportunities. Businesses seek these opportunities to remain competitive and to realize profits from sales. In its licensing and patenting services the Office ensures that the commercial process is expedited and that the business licensee is adequately protected under patent law to enjoy the rewards promised by the venture.

The Public

The public ultimately benefits from TTO's marketing strategy of exposing its inventions to a broad range of businesses. Such exposure increases the likelihood that the most promising new inventions will be matched with the businesses most able to bring them to the market. Economic growth and technological progress are thereby enhanced. The public is also served by the TTO's patenting program, since its patents are available as teaching documents.

Besides being available to the public as educational instruments, the presence of patents in the public literature promotes further research in the subject of the patent. Such research can lead to the continued development and enhancement of knowledge in the field. Breakthrough patents often give rise to increased R&D spending in the respective fields. That often leads to additional patents and additional licenses, which in turn lead to increased commercialization, economic growth, and public benefit.

OPERATIONAL FUNCTION

This function has responsibility for carrying out the required steps in the Patent Policy that are intended to ensure that inventions are disclosed and that a public benefit is realized as a result. They are listed below in the order in which they are generally accomplished.

1. Identifying inventions
2. Marketing and securing research support
3. Making licensing and other arrangements
4. Securing appropriate patents
5. Managing financial arrangements (managing patent budgets, royalties, and other invention-related income.

While discussed separately below, the responsibilities are generally interdependent.

Identifying Inventions

In the TTO, the identification of an invention occurs after it receives an invention disclosure. It refers to the steps the

Office takes to determine the respective property rights that would accompany the invention. As an inducement to inventors to disclose their inventions, the TTO relies primarily on its continuing delivery of quality services to inventors and on its resulting reputation as a service provider. Furthermore, it relies on the financial incentive for disclosure provided under the Patent Policy.

Once the Office receives an invention disclosure, it initiates communications with the inventor(s), the campus where the invention originated, and the research sponsor(s). It reviews the circumstances of the invention, the interests of individual and institutional participants, and the place(s) of the invention-related activity. Also, it takes into account terms and conditions of applicable grants and contracts. The process clarifies ownership rights and obligations to sponsors, if any. The Office also determines whether or not the disclosure constitutes an invention that would convey property rights to any or all of the participants in the research. Its guide in this regard is the criteria of patentability under patent law.

Marketing and Securing Research Support

The success of a venture often depends on a long-term collaboration between a prospective licensee and the inventor. Before the Office markets the invention to third parties, therefore, it must take into account certain situations. One is whether the inventor already has a company with whom he or she wishes to work. Or the inventor may wish to use the technology as a basis for starting up his or her own company. Another factor is that the invention may be a candidate for incubation under a local economic development endeavor.

In the absence of a preferred licensee candidate ready to accept the invention for commercialization, the TTO exposes the invention widely among licensing candidates. The greater the exposure of an invention, the higher the likelihood that it will eventually find its place in the market. In like manner, an invention is most likely to capture its best development opportunity if all the offers from qualified licensee candidates are considered. Besides established firms in the field, licensee candidates include venture capitalists and entrepreneurs. The latter play an important role in new company formation.

The Office is committed to informing the industrial sector of all available inventions, and the exposure of the new technologies is specific to each field. An important corollary activity is the exposure of faculty inventors to the interests and requirements of the market and to the opportunities to develop industrial linkages that such exposure provides. As with licensing, industrial research support is often the result of inventors knowing people and being known by research and development people in industry. TTO conducts its marketing program in a manner that maximizes the opportunities for these linkages.

Making Licensing and Other Arrangements

It is not uncommon during the marketing process for the Office to negotiate and execute a number of different types of agreements. At these earlier stages, the likelihood of the discontinuation of interest in an invention is high. The new technology evaluation agreement enables licensing candidates to obtain full access to new technologies for evaluation and testing purposes while protecting patent or proprietary rights. Forms of this agreement include confidentiality agreements, testing agreements, screening agreements, materials agreements, and variations thereof.

Option agreements grant certain rights in situations where a higher level of interest is apparent. The option agreement is also an evaluation opportunity, granted for a limited period of time, in return for compensation. Generally, it grants a company the first right to negotiate a license on the subject invention and/or on additional inventions arising from work relating to the invention. The option grant may be a stand-alone agreement or it may be a provision in a research contract, testing agreement, screening agreement, or materials agreement.

The actual scope, terms, and conditions of a commercialization arrangement are spelled out in the license agreement. The areas negotiated by the TTO and a company include the technology covered, milestones for the development of the invention, royalty terms, reporting obligations, patenting and actions to be taken in the event of infringement, and others. These areas have been addressed in previous chapters. Additional considerations are involved in the event of equity holdings and control of the licensed company by an inventor and/or the Foundation.

The TTO works with campus officials to ensure that the executed license does not give rise to conflicts of interest. Here, the parties are guided by SUNY's Conflict of Interest Policy. This policy applies to all activities involving the private business interests of University employees as they relate to use of University facilities. Other important guidelines in the licensing process concern Equity Participation, Emerging Technologies, and Cooperative Use of Research Equipment.

Securing Appropriate Patents

The process of determining which inventions should be patented is selective, since the patent budget does not allow for the patenting of all inventions. The Office deals daily with inventors, with individual campuses advocating the particular merits of their individual technologies, and with recurring states of urgency. The imminence of a public disclosure or the approach of a statutory bar terminating the grace period for filing a patent application are two examples.

Generally, there is no problem in filing a patent application if an industrial sponsor willing to reimburse patenting costs has been located. For filings on licensed technologies with no direct cost reimbursement, determining factors include the requirements of the technology and the promise of the market, that is, whether or not the expected royalty warrants the cost of the application process.

Subject to budget constraints, patenting may also be pursued before a licensee is located, and feedback from marketing is important in these determinations. Generally, high value is placed on technologies representing fundamental discoveries. These are regarded as seminal inventions because they hold promise to be the basis for further invention and patents. Inventions with broad applicability or potentially multiple applications are also favorites, as are those perceived by the Office to be technologies of the future, even though they arouse no immediate market interest.

Managing Financial Arrangements

The TTO maintains administrative oversight over all the obligations under the license agreement. It manages the accounting of royalties received and their distribution to the appropriate people. It handles bills for patent expense reimbursement and closely monitors filing and maintenance costs. Generally, business activities with royalty implications occur under the Royalty, Due Diligence, and Reports Articles. Sometimes when there is a termination fee, the Termination Article applies. Due Diligence and development milestones set the pace of progress toward market introduction, dating sales activity. The reporting obligations in the agreement require the licensee to communicate whether or not the milestones are being met or whether or not royalties are being earned. The Office uses the reports from licensees, or their absence to schedule in-house performance reviews of licenses and conducts the appropriate follow-up.

COMPLIANCE FUNCTION

Compliance with Sponsor Regulations

The Bayh-Dole Act, under which institutions receiving government grants may acquire ownership rights to government-sponsored inventions, requires that patenting decisions for such inventions be made in a timely fashion. In addition, the institution is expected to actively seek commercial sponsors to ensure market entry for the new product or process. Failure to do so may result in forfeiture of rights in the technology.

Companies, philanthropic organizations, and various types of other nonprofit and public institutions also sponsor research. In the event of new technology disclosure, TTO reviews the respective grant and/or contract terms governing inventions and patents and acts to satisfy those obligations that apply. The Office also acts to satisfy the requirements of the Patent Policy regarding inventors' rights relating to participation in royalties and reacquisition of rights in the event of nonpursuit of commercialization or patents.

The Foundation

The TTO is both executor and custodian of SUNY- and Foundation-owned intellectual property as well as the resulting agreements and income relating thereto. In its executor role, the Office discharges its marketing, patenting, and licensing functions described under the Operational Function. In its role as custodian, the Office ensures a full accounting of all intellectual property and related assets. They include the disclosed new technologies, patent applications, patents, and the various commercialization agreements that relate to them. Table 1 below is illustrative. The data in the table are fictional.

Table 1
Consolidated Statement of
Foundation and State University of New York
Active Intellectual Property Holdings and Related Assets

For the period ending December 31, 1995

Intellectual Property/ Assets (Agreements)	Total at 01/01/95	Additions	Deletions	Total at 12/31/95
PORTFOLIO DATA (#)				
Disclosures	300	50	45	305
Federal	100	30	15	115
Nonfederal	50	10	15	45
SUNY	50	7	10	47
Software	75	2	3	74
Incomplete	25	1	2	24
Patent Applications	70	20	15	75
United States	35	10	7	38
Convention Filings	5	2	3	4
Foreign — National	30	8	5	33
# of Inventions	60	20	15	65
Patents	300	60	55	305
United States	200	40	35	205
Foreign	100	20	20	100
# of Inventions	250	55	50	255
Licenses & Options	200	50	45	205
Exclusive	100	20	20	100
Nonexclusive	90	20	20	90
Options	10	10	5	15
Other Agreements	650	200	180	670
Confidentiality	520	150	120	550
Materials	80	30	45	65
Screening	30	15	9	36
Institutional	20	5	6	19
Technology Source				
Number of Inventors	200	60	50	210
Number of Campuses	15	12	9	18
FINANCIAL DATA ($,000)				
Income and Expenses				
Royalties	$6,000k	300k	200k	$6,300k
Reimbursements	2,000k	200k	150k	2,200k
Patenting Costs	2,500k	300k	250k	2,800k

This Consolidated Statement serves a number of purposes. It shows the cumulative active stock of intellectual property and related assets at a particular point in time, and it shows the change in these holdings from one period to another. The two middle columns, "Additions" and "Deletions," report the activity on the portfolio during the intervening period. This may be a month, quarter, or year, depending on whether the statement is required as an inventory report or as an operations monitoring instrument. The magnitudes of Table 1 also reflect workload to a degree, and they can be used in the development of indices of productivity, efficiency, cost effectiveness, and return on investment.

Disclosures — he first category reports Disclosures. These are the inventions and software under TTO management. Because different types of disclosures give rise to different types of obligations to sponsors, relating to the patenting, licensing, and reporting of the new technologies, data under Disclosures are further broken down into subcategories. Since sponsor obligations generally do not become active unless there has been a complete disclosure of an invention, those the Office receives as "Incompletes" are categorized separately and receive a different kind of administrative attention.

Patent Applications — Patent applications are complex and legal services are provided by outside patent attorneys. For domestic applications, the TTO administers and keeps a watchful eye over the filings, prosecution, Patent Office actions, and various transformations the applications may undergo, such as into CIP's, Divisionals, appeals and sometimes the complications of interference proceedings.

For the most part, the TTO renders these services for foreign filings as well. However, in foreign filings there is the additional requirement of filing under the various international treaties and the subsequent filings in the individual nations. During this process, United States patents may be issued or applications may be discontinued or abandoned, giving rise to data for the "Deletions" column.

Patents — The Office also monitors the life of issued patents. A patent bestows on its owner a right to legal recourse in the event of unauthorized use of the patented invention by someone else. The patent represents legal property, acquired through a very costly applications process, and it acknowledges the patent holder's intention of realizing financial reward through commercialization of the invention. A report on patents is the inventory of these rights. Patents must also be maintained to remain active, and the TTO must monitor and make payment of the appropriate fees.

Licenses and Options — The charge of the Office is to locate businesses that will enter into license agreements so that the public can ultimately benefit from the new technology through its commercialization. The option is a significant step in that direction. Numbers for licenses executed are indicative of the technologies actually developed and being developed by the industrial sector into new products and processes. Often several inventions are incorporated under one license, depending on interrelationships or the potentialities of such combinations. Post-agreement follow-up by the TTO includes administration of royalties and monitoring of the compliance of licensees with the terms and conditions of the agreements, such as due diligence and reporting obligations.

Evaluation Agreements — This category includes what are generally considered to be technology evaluation agreements, in particular, those that usually do not involve remuneration or only in relatively small amounts. They contain time frames and are used to protect proprietary information or developments arising therefrom. Because they represent contractual obligations, the TTO must be vigilant about monitoring the performance thereunder, the outcome of performance, and the implications for the technology after the contracts expire.

Technology Source — The inventors served by the Foundation are located at a number of different campuses. The faculty and campus officials at each campus are responsible for the research endeavors, outcomes, and disclosure of new technologies at their own campuses. More recently, they have assumed decision-making authority in the patenting process. The campus officials are responsible for reporting to the campus president. In working with the TTO, the contact person for the campus is usually the head of the campus sponsored-funds office (Sponsored Funds Officer or SFO).

The TTO maintains close communication with each inventor as management of the case proceeds, that is, through the disclosure, sponsor reporting, marketing, patenting, licensing, and royalty administration phases. Copies of all correspondence to the inventor are sent to the SFO. Such itemized contacts with the SFO, however, are not a helpful indicator of TTO services rendered to all the inventors on that campus or of the overall benefit obtained by the campus. To that end, the TTO performs a separate reporting.

Income and Expenses (in thousands of dollars) — The Totals reported here are cumulative, showing amounts received and expended during the life of the Office. Accordingly, if the interval of the Statement is one year, then the intervening columns will show first the current year's amount, then the previous year's amount. In this case, the "Beginning Total" is increased by the current year's total to arrive at the "Ending Total." These figures are associated with the cumulative active totals reported under the preceding titles. For example, a royalty receipt in the current year may be from a license that has been active for 10 years, or a patenting cost may relate to an application that has been pending for several years. However, since they are lifetime figures, the cumulative amounts include monetary transactions on expired licenses and patents. An annual average is obtained if the cumulative Totals are divided by the number of years they represent, thereby providing a useful comparison for annual amounts in the middle columns.

Some of the tasks requiring TTO follow-up are as follows. For royalty distribution to occur, a royalty distribution agreement must be executed by the inventors. The inventors are also required to complete certain forms for income tax purposes. The tasks become more complex in the event of deceased inventors or bankruptcy. Patent Cost Reimbursement entails billing of companies and maintenance of payment records. It also requires close oversight of attorney charges and expense authorizations so that such charges stay within the necessary budget constraints.

COST SHARING AND RELATIONSHIP OF SERVICES

The colleges and universities enlisting the services of the Office realize significant savings in management and administrative costs, resulting mainly from economies of scale. Smaller institutions enjoy technology transfer services that would otherwise be beyond their reach. Those with adequate budgets receive such services at relatively lower cost.

Ideally, an invention progresses from disclosure to commercial interest, to license, to patent, in that order. Rarely is a patent application filed on an invention that is unable to elicit immediate or future market interest. In the absence of a licensee, the potential business interest revealed through marketing guides the patenting decision. This response from the industrial sector has an influence on the negotiation process when a candidate is located. It affects royalty expectations, and the terms and conditions of the license agreement. The response also determines whether or not the Office will retain an invention, return it to the inventor, or relinquish it to the sponsor.

Notes

[1] The organization of the material and interpretation of the Foundation's Strategic Plan and policy objectives presented in this chapter represent the opinion of the author.

[2] Inventions and Patent Policy of State University of New York.

[3] Ibid.

[4] Strategic Plan of The Research Foundation of State University of New York, 1996.

Chapter 11

Technology Transfer Office Performance Evaluation

Generally, the performance of an entity involves the execution of many functions, and analysts have developed the concept of an index to evaluate performances that involve such multiple inputs. Consider the Index of Leading Economic Indicators, the Dow Industrial Average, or the Consumer Price Index. All of these indexes are used widely, and exert powerful influence on decision makers.

To evaluate TTO performance, a Technology Transfer Office Performance Index (TTOP Index) is proposed. It is a single, composite number that takes into account licenses, patents, and the financial values they embody as well as the processes by which these values are achieved. The Index can be calculated periodically, and comparisons of one period with another can be used to highlight strengths and weaknesses in services.[1] Adjustments to the index may be made, to take account of the relative importance of each of its components.

ECONOMIC INDICATORS

The amount of money generated by the relationships the TTO creates is probably the most popular performance indicator in use. Royalty income has the advantage of simplicity, and is readily understood by everyone. Research support received from licensing is also sometimes viewed as a performance indicator.

Both royalty income and research support arising out of licenses depend substantially on the market relevance of invention disclosures and exploitation of already licensed technology in the commercial sector, events over which the TTO may have little control. These data nonetheless are significant determinants of institutional support for the Office.

The data provide a route to measuring the economic impact of technology transfer, the public benefit sought by institutions when they create TTO's. Therefore, the ability of the Office to demonstrate this outcome cannot be overlooked. Fortunately, available techniques in economic analysis may be utilized to describe this very important result of technology transfer.

An example is the technique associated with the Nobel Laureate in economics, Wassily Leontief. The model recognizes that differing technological needs cause industries to respond differently to stimuli. Dynamic interactions involving them as employers and buyers and seller to each other and the consuming public cause a change in demand for the products of one industry to have ripple effects throughout the economy. The result is a chain reaction causing an eventual level of output within a year which is several times the amount initiating the change.

The input-output method, or interindustry analysis as the model is known may be used to measure local, regional, or national economic impact, depending on the scope of the analysis desired. One study utilized the model to calculate the impact of federal expenditures for academic R&D on the economy of New York State.

The model estimated the resulting output from new spending at 3.7 times the level of the initial R&D amount. This produced a sufficient national and state tax collection to offset the federal expenditure.[2] To begin, this expenditure could be new industrial R&D spending and other costs relating to the exploitation of the licensed technology. Such spending creates additional demand for inputs and services, thereby raising output and employment levels.

PERFORMANCE INDICATORS OF THE TTOP INDEX

Research dollars and royalty receipts figure prominently in Office output and in the TTOP Index. However, institutional statistics across the nation show few with licensing rates more than thirty percent of inventions disclosed. Also, royalty income is often highly concentrated in a few inventions. Therefore, sole use of these amounts would not be indicative of performance on the range of services demanded by the typical inventions portfolio.

The TTOP Index takes account of both licensed and yet-to-be licensed technologies. In the former instance, the promise of income occurs under the license. Regarding the latter, the TTO uses the disclosure to seek licensees, initiating communications between the respective researcher inventors and their scientific counterparts in industry in the process.

The criteria are judgments of the author, and all the values assigned, the measurements and the importance weights are hypothetical. In reality, these determinations should reflect the priorities and concerns of the institution, and its designated decision makers. The process might involve nomination of an expert group or committee. Each member thereof could develop a list of criteria. Results could then be pooled and areas of agreement identified. Finally, agreement must be reached on importance weights, for which the same process can be employed.

The TTOP Index therefore, also takes account of both quantity, indicative of effort, and quality of service. It rests on the simple understanding that performance is better or worse if it is better or worse than previous performance. As will become evident, few in technology transfer would doubt that the most favored direction for the individual performance measures, and consequently the TTOP Index, is upward.

The Index uses five broad performance measures, or indicators. They are,

1. Invention Disclosures
2. Company Evaluations of Inventions
3. Income-Generating and Industrial R&D Support Agreements
4. Patentability Opinions, Patent Applications, and Issued Patents
5. Institutional Support for the TTO.

Invention Disclosures

Institution researchers are the developers or suppliers of invention disclosures. If we assume that they behave as suppliers, a demonstration by the TTO of remunerative services and rewards should increase their supply of inventions. A reward is usually financial return through licensing and royalty income or recognition by appearance on an issued patent as a named inventor.

Service is the effort expended to accomplish these ends. The TTO must exercise due diligence in these efforts, and it may communicate this due diligence under the following headings.

(a) Licensed technologies — communications with the inventors in regard to progress and status of licenses, income and patent rights
(b) Unlicensed active cases — communications with the inventors concerning progress toward commercialization agreement(s), reactions of licensing candidates, and patenting
(c) Inactive cases — communications with the inventors indicating reasons TTO has decided to relinquish rights, including documentation of marketing efforts, names of companies contacted, known reasons for TTO's failure to find an industrial sponsor, and patenting possibilities.

Whether or not an invention is ultimately licensed or patented, the TTO's individualized attention to inventors will encourage them to disclose subsequent inventions and will encourage disclosures from new inventors as efforts to achieve rewards and/or establish linkages between inventing researcher and the industrial community become known. Accordingly, the measure of performance for the Invention Disclosures indicator is the ratio of total disclosures to total researchers and the ratio of disclosures to the total number of R&D projects. The number of departments participating is also important. Therefore, the TTOP Index includes the relationship between disclosing departments and the total number of departments.

Company Evaluations of Inventions

Industrial liaisons and the creation of linkages between inventing researchers and the industrial R&D community constitutes what is typically referred to as technology marketing. This linkage, whether or not the direct result of TTO liaison, ultimately determines the invention's fate as a new product/process candidate.

For the reviewing company, an evaluation may result in the selection of one invention among many competitors for licensing and commercial exploitation. The learning experience is important in this process. The company becomes aware of the commercial and technological strengths and weaknesses of the evaluated technologies and gets to know the respective investigators as well as the capabilities of the originating institutions. It acquires an appreciation of the originators' usefulness as potential partners in R&D. This latter point has significant implications for immediate and future relationships.

Inventors learn from this institution/industry dialogue also. Typically, a comprehensive corporate technology review includes extensive direct conversations between the inventing researcher and industrial counterparts. The inventor gets direct technical and market-related feedback on business needs and can acquire an improved awareness of the market relevance of his or her technology and continuing research.

He or she may achieve a relationship with the industrial community, not unlike that achieved with scientific colleagues as a result of publications in scientific journals and presentations at professional conferences. These outcomes can be beneficial in both the short term and the long run in the potential they have for increasing the licensibility of research results. Inventors and institutions may lose these valuable opportunities when the TTO does not actively market inventions.

For the present indicator, the concept of Evaluation Agreement is defined. A comprehensive discussion of proprietary information with third parties is generally not recommended unless an agreement is in place to protect confidentiality. The Evaluation Agreement serves this purpose and includes all the agreements that protect the confidentiality of a technology while a licensing candidate obtains information or samples in order to fully understand the invention. The Evaluation Agreement marks the beginning of comprehensive technical discussions with the inventor. The number of Evaluation Agreements reached is thus a valuable indicator of the number of direct industrial connections established by the TTO.

The Evaluation Agreement may be a Confidentiality Agreement, Testing Agreement, Screening Agreement, Materials Agreement, or Options Agreement. The relationship of these agreements to the number of companies contacted and the number of inventions represents the effort expended on marketing and the quality of that effort.

Campus visits by companies and visits by inventors to company sites also promote researcher inventor/industry ties and further enhance opportunities for industrial R&D support and commercialization of technologies. These happenings are captured in the TTOP Index as desirable outcomes of marketing.

Income Generating and Industrial R&D Support Agreements

The TTO has an obligation to the institution and the inventors who share in royalties to ensure the best returns on licenses and other income-bearing agreements. The TTO must be especially thorough in

(a) Background preparation such as market analyses, valuations of the technology, and determinations of the technical and financial adequacy of licensee(s) prior to negotiations;
(b) Oversight and monitoring of executed technology transfer agreements to ensure optimal recovery of returns.

Greater numbers of such agreements relative to number of disclosures reflects TTO effort expended. This ratio is also indicative of participation among inventing researchers in actual and potential royalty streams. When the income generated is examined, its relative amount is an important consideration, and the TTOP Index includes it as a ratio of the total institutional budget.

The ability of the TTO to bring in industrial support, directly and indirectly, is included as a ratio of industrial R&D to total R&D sponsored at the institution. As noted earlier, recognition by industry of institutional and researcher strengths can lead to relationships in which industries draw on them as resources in industrial new product/process research and development. The TTO's role as liaison is important in establishing this recognition, and the number of actual agreements made indicates the success of the TTO's efforts.

Patentability Opinions, Patent Applications, and Issued Patents

An invention is usually an obvious candidate for being patented if it receives a positive patentability opinion and attracts a potential licensee who is ready and willing to pay for a license. However, the potential licensee is not always present. Also, filing for a patent might not be the best option, even if the patentability opinion is positive. Know-how licenses, licenses not involving patents, are not uncommon.

With the above considerations in mind, the task for the TTO is to allocate its spending so as to achieve the broadest satisfaction among the inventing researchers in terms of allocations for patentability opinions and patent applications. At the same time, it must achieve maximum value for its patents portfolio. The TTOP Index measures this effort by the relationship of patentability opinions and patent applications filed to invention disclosures. The number of options and licenses relative to issued patents and the number of issued patents relative to patent applications combine to define quality in the patenting process.

Institutional Support for the TTO

The institutional budget must respond to many competing interests. Departments and divisions are cut from the budget in this competition, which is most severe in times of fiscal austerity. At many institutions, the TTO venture is a relatively new undertaking and therefore vulnerable when competing with established departments. Consequently, funding will be at even greater risk if considerations of value and good performance are not effectively presented to the top institutional administrators and used to make the case for technology transfer.

Due to its impact on the budgeting process, the TTO services provided to enhance this indicator have implications for all the service categories indicated. Services might include comprehensive top-level management reports and TTO advocacy at Board of Directors' meetings. The TTO's effectiveness in representing technology transfer interests and needs appears in the TTOP Index as a ratio of the TTO budget to the total institutional budget.

CALCULATING THE TTOP INDEX

Calculation of the TTOP Index is illustrated in Table 1. First a base period is determined. In this example, it is the 5 years ending June 30, 1990. Then ratios for the new years are calculated and expressed as percentages of the base period averages. Data for 1991 appear first, then data for 1992, then for 1993. All data are hypothetical.

In each year, an average is calculated for each indicator. This is obtained by simply adding the percentages reported and dividing by the number of such percentages; for example, the 107 indicated as the 1991 Average of Disclosures is arrived at as follows: [(115 + 100 + 105) = 320], then 320/3 = 107. The TTOP Index is obtained by averaging the averages, as follows for 1991: [(107 + 123 + 102 + 96 + 85) = 513], then 513/5 = 103.

The Index may be used to display total performance over time. Variations in the indicators can be charted also to show the areas of activity contributing to variations. Separate detailed analyses of the underlying causes of variation in the individual performance measures can then be conducted to help explain the observed overall annual differences in the Index.

For example, in the data of Table 1, the 1991-92 Index gain is attributable to increases in the Agreements, Patenting, and Support of TTO indicators. Declines are reported for Disclosures and Evaluations. However, Disclosures contribute significantly to the growth of the Index in 1993, and a major portion of this growth is traceable to the inventions/researcher ratio.

By expressing the base-year performance relationships

Table 1
Calculation of Technology Transfer Office Performance Index

Performance Indicator Relationships (Overall Totals)	Base Average	% of Base Average 1991	1992	1993
# Inventions/# Researchers	.15	115	130	150
# Inventions/# R&D Projects	.10	100	80	110
# Disclosing Depts./# Total Depts.	.25	105	100	115
Average for Disclosures		107	103	125
# Evaluation Agreements/# Contracts	.05	120	110	125
# Evaluation Agreements/# Inventions	.85	140	120	130
# Company Visits/# Evaluation Agr.	.05	110	130	120
Average for Evaluations		123	130	125
# Options & Licenses/# Inventions	.20	90	105	120
$ Royalty/$ Institution Budget	.30	110	95	115
# Industrial R&D/# Total R&D	.45	105	120	125
Average for Agreements		102	107	120
# Patent Opinions/# Inventions	.65	108	80	100
# Patent Applications/# Inventions	.18	106	130	140
# Options & Licenses/# Patents	.10	70	120	140
# Patents/# Patent Applications	.75	100	120	140
Average for Patenting Activity		96	113	130
$ TTO Budget/$ Institution Budget	.12	85	112	115
Average for Support		85	112	115
TTOP Index		103	111	123

as ratios, otherwise incomparable data can be compared, and by using averages in calculating the Index, benefits deriving from size alone are factored out. As a comparative metric, the Index enables comparison of large and small institutions. For those who question the relevance of certain indicators or components thereof, the TTOP Index is amenable to substitutions, additions, and deletions; abnormal fluctuations are contained as well.

ADJUSTING THE INDEX FOR THE RELATIVE IMPORTANCE OF INDICATORS

This section responds to the concerns of those who argue that certain of the indicators are more important than others. It incorporates importance weights in the Index to take into account relative significance. The approach recognizes a hierarchy by first scoring each indicator, then each criterion. A weighted average becomes the basis for the new index (Index-II, the objective sought by the calculations is ranked as LEVEL 1 and is reported in Table 3).

To arrive at normalized weights, indicators are first ranked according to importance. Then an importance weight is assigned to each, with no restrictions. For example, an indicator with a score of 20 is twice as important as an indicator with a score of 10. Finally, a calculation of the sum of the weights is made. The answer provides the denominator, which is then divided into the importance weight of each indicator. The sum of the resulting calculations equals "1." A similar process is then undertaken for each criterion of the respective indicators.[3] Table 2 is illustrative. The assignment of importance weights may be read as follows. Among the Indicators, Agreements are rated as twice as important as patenting, with weights of 20 and 10, respectively. With a score of 2, Support for the Office is the least important in determining TTO performance. The weights are normalized by summing the weights for the Indicators and dividing this total into the importance weight of each Indicator. The total is 44. Each score is divided by this total, giving the assignments in parentheses (e.g., 20/44 = .45, etc.).

Next, performance within each Indicator is read by looking at the individual performance criteria. In the Agreements Indicator, the ratio of royalty income to institutional budget ranks highest, with a score of 6 times that for the ratio of industrial R&D to total R&D; scores of 30 and 5, respectively. The total of the criteria score for Agreements is 45 (i.e. 30+10+5=45). These weights are also normalized, within each Indicator. With 45 as the denominator for Agreements,

Table 2
Criteria Ranking and Calculations of the Weighted Index

LEVEL OF IMPORTANCE INDICATORS CRITERIA		
LEVEL 2	LEVEL 3	WEIGHT
Agreements 20 (.45)	$ Royalty/$ Institution Budget 30 (.67)	.30
	Options & Licenses/# Inventions 10 (.22)	.10
	# Industrial R&D/# Total R&D 5 (.11)	.05
	Subtotal	.45
Patenting 10 (.23)	# Patents/# Patent Applications 50 (.42)	.10
	# Options & Licenses/# Patents 40 (.33)	.07
	# Patent Applications/# Inventions 20 (.17)	.04
	# Patent Opinions/# Inventions 10 (.08)	.02
	Subtotal	.23
Evaluations 8 (.18)	# Evaluation Agts./# Inventions 50 (.50)	.09
	# Evaluation Agts./# Contacts 40 (.40)	.07
	# Company Visits/# Evaluation Agts. 10 (.10)	.02
	Subtotal	.18
Disclosures 4 (.09)	# Inventions/# Researchers 60 (.50)	.05
	# Inventions/# R&D Projects 50 (.42)	.03
	# Disclosing Depts./# Total Depts. 10 (.08)	.01
	Subtotal	.09
Support 2 (.05)	$ TTO Budget/$ Institution Budget 100 (1.00)	.05
	Subtotal	.05
	Total of Importance Weights	1.0

the weight of 30 calculates to an assigned normalized weight .67 (i.e., 30/45=.67), in parenthesis.

The final weights for each Indicator are obtained by multiplying through the levels. For example, Agreements has a final weight of the sum of the weights for $Royalty/ $Institution Budget, #Options & Licenses/#Inven-tions, and $Industrial R&D/$Total R&D:

LEVEL 2 (.45) x LEVEL 3 (.67) = .30
LEVEL 2 (.45) x LEVEL 3 (.22) = .10
LEVEL 2 (.45) x LEVEL 3 (.11) = .05

Totaling these gives the total of importance weights for all the Indicators, and this equals "1."

The final weights are then applied to the raw data on each criterion variable. This is performed using data from Table 1, in which the unadjusted TTOP Index is calculated. The process of arriving at the adjusted Index, Index-II, is illustrated in Table 3. The weights are simply applied to the percent of the base-period average for each criterion. The result is summed for each Indicator. The total for the Indicators (which is the same as the total for the criteria) provides the new adjusted TTOP Index-II (i.e. 117 for 1996).

The aggregation formula for any one period is as follows: TTOP Index-II = $\Sigma[u(ij)100/b(i)][w(i)]$, where u(ij) is the value of criterion i in period j; b(i) is the base average of criterion i; the result of each such calculation is multiplied by 100, and the resulting product is multiplied by w(i), the weight assigned to criterion i. Below is an example of one iteration, where i is the criterion #Industrial R&D/#Total R&D: #Industrial R&D/#Total R&D = [.52(100)/.45][.05] = 5.77, rounded to 6. In 1996, the j year, the ratio of #Industrial R&D to #Total R&D equaled .52.

Table 3
Calculation of Technology Transfer Office Performance Index-II

Performance Indicator Ratios/Criteria	Base Average	% of Base/ 1996 (Weight)	Weighted 1996 Index
$ Royalty/$ Institution Budget	.30	130 (.30)	39
# Options & Licenses/# Inventions	.20	110 (.10)	11
# Industrial R&D/# Total R&D	.45	115 (.05)	6
Total for Agreements			**56**
# Patents/# Patent Applications	.75	105 (.10)	11
# Options & Licenses/# Patents	.10	120 (.07)	8
# Patent Applications/# Inventions	.18	95 (.04)	4
# Patent Opinions/# Inventions	.65	80 (.02)	2
Total for Patenting Activity			**25**
#Evaluation Agreements/# Inventions	.85	108 (.09)	10
# Evaluation Agreements/# Contacts	.05	120 (.07)	8
# Company Visits/# Evaluation Agr.	.05	80 (.02)	2
Total for Evaluations			**20**
#Inventions/# Researchers	.15	90 (.05)	5
#Inventions/# R&D Projects	.10	100 (.03)	3
# Disclosing Depts./# Totals Depts.	.25	80 (.01)	1
Total for Disclosures			**9**
$ TTO Budget/$ Institution Budget	.12	110 (.05)	6
Total for Support			**6**
TTOP Index-II			117

To conclude, the Index recognizes services rendered on all inventions. The TTO is asked to maximize patents, royalty income, and research support, as well as account for its role in using invention disclosures to create industrial linkages for researcher whose disclosed technologies show no immediate market interest. In its latter role, the TTO seeks to enhance the number of market-relevant inventions, and therefore licensable and patentable inventions, produced in the future by promoting awareness of industrial needs among researchers. This dual approach ensures maximization of invention disclosure potentialities in both the short term and the long run.

The methodology used is adaptable to the needs of practitioners not entirely satisfied with the particular criteria, indicators, or weighting system proposed. It is flexible as to substitution and recognition of differing perceptions of importance of the indicators. Not only is the Index useful for single office applications, it provides an approach for aggregate representations of TTO performance, in which data from institutions across the nation may be combined in a single National TTOP INDEX (NTI).

Notes

[1] Albert E. Muir, "Technology Transfer Office Performance Index," *Journal of the Association of University Technology Managers*, Vol. V (1993): 61-73.

[2] __________, "Interindustry Analysis of the Impact of Federal Support for Academic Science on the Economy of New York State", *Research in Higher Education*, Vol. 18, No. 2, (1983): 412-433

[3] A. W. McEachern, "Two Simple Versions of Multiattribute Utility Analysis," in *Decision Making In the Public Sector*. Ed. by Lloyd G. Nigro (New York: Marcel Decker, Inc., 1984), 65-77.

Chapter 12

Dimensions of Public Policy

Government, business, independent inventors, universities, and nonprofit institutions all play important roles in technology transfer. The federal government, however, is singular in terms of the overall implications of its role. In the United States, we expect government to provide and protect the setting for technological progress, and thereby for technology transfer.

> The development and diffusion of advanced technologies requires a system of education and training as a basis for supplying technology and skills, a legal framework for defining and enforcing property rights, and processes . . . to reduce transaction costs and increase the efficiency and transparency of markets.[1]

The federal technology transfer policy issues from a mix of policies, laws, and regulations, designed to foster the development and diffusion of advanced technologies. The government also plays a major a role in the nation's research and patenting enterprise. This is a primary source of the inventions for technology transfer. It is both a source and custodian of intellectual property rights. Furthermore, participation of the federal government in technology transfer affects the way other parties in the process, in particular prospective licensors and licensees, behave. Antitrust laws guard against corporate conduct and behavior that pose threats to innovation. The ultimate purpose of policy is to induce technological development, and the effects of this are evident in indicators of progress.

GOVERNMENT RESEARCH AND PATENTS

Technology transfer is related to the nations science and technology policy. The federal government's science policy drives the nation's R&D program, supporting research and development programs at universities and in business, government laboratories, and other private and nonprofit institutions. Its technology policy is designed to protect the security and health of the nation, U.S. competitiveness in world markets, and international trade flows.

Government technology policies can be characterized as either mission oriented or diffusion oriented. Mission-oriented policy seeks to achieve clearly defined national goals. It is characterized by a high degree of concentration of funding in specialized fields, such as those most likely to bolster national defense. Diffusion-oriented policy seeks to achieve as broad a direct public benefit as possible. Federal policy is a mix of the two. It is predominantly mission oriented in funding research and predominantly diffusion oriented in promoting technology transfer.

Research Support

Government is both a sponsor and a performer of research. In its role as sponsor, it supports research at universities, nonprofit corporations, and in industry. Virtually all federal agencies conduct research. However, most of the federal research money is directed to agencies concerned with national security: the Department of Defense (DOD), the National Aeronautic and Space Administration (NASA), and the Department of Energy (DOE). Thus in 1990, 63 percent of federal funds for industrial R&D was dedicated to aircraft and missiles.[2] The Public Health Service receives slightly over 10 percent, and its funding focuses on biomedical research. The National Science Foundation (NSF), concerned with non-mission-oriented research, receives less than 3 percent.

Patenting Activity

The technology transfer potential of government research is best illustrated by its statistics on patenting activity. Potential is the operative measurement, since most of these patents, while available for licensing, do not actually find commercial sponsors. During 1993, the number of U.S. patents granted to the federal government exceeded the number granted to any corporation, either domestic or foreign.[3] By so protecting the results of its research, the government is able to withstand property right challenges, either as a plaintiff or as a defendant, as it promotes or practices the technology it develops in the public interest.

TITLE AND RIGHTS TO INVENTIONS

The present federal policy on technology transfer is possible largely because of the nation's patent and trade secret laws. While the latter fall under the jurisdiction of the states, ownership rights may be generalized for the nation. Recent legislation with direct consequences for technology transfer includes the University and Small Business Patent Procedures Act (the Bayh-Dole Act) of 1980 (P.L. 96-517), the Stevenson-Wydler Technological Innovation Act of 1980 (P.L. 96-480), and subsequent supplemental legislation. These Acts determine what happens to publicly sponsored inventions, and their implications are widespread given the scope of the federal research enterprise.

Most inventors today act in their capacity as employees in the private and public sectors. Under the common law doctrine, the employer owns the invention if the invention is the result of work-for-hire and if the invention is within the scope of the employee's responsibilities. In addition, under the shop-right doctrine, the employer receives an unassignable, nonexclusive, royalty-free license to use an employee's invention, provided that company time, facilities, and material were used.

Often companies negotiate specific contractual obligations with their employees in regard to inventions, patents, and know-how. In particular, they require the employee to assign inventions, (to transfer the new technology to the company), cooperate for purposes of patenting, and refrain from disclosing confidential information and trade secrets. Rights go to the inventor if the invention is made during his or her own time, using his or her own facilities, just as they would to an independent inventor.

In the case of a government, university, or nonprofit employer, the employee's rights are generally subject to unilateral rules, regulations, and statutes. Under the rules, ownership rights are automatically assigned (the invention is transferred) to the employer as a condition of employment. Usually the employer retains rights to all inventions except those made during the employee's private time at his or her own facilities.

Many individuals may be involved as contributors to a discovery in the modern laboratory, and individual contributions are not always obvious. The collaboration of the inventors is also important in the exercise of rights in new technologies as well as in the accomplishment of research and development missions. By giving outright assignment to the employer, it is believed, the path is clear for the employer to exploit the invention commercially either as a producer or as a licensor, as in the case of nonprofit entities.

The federal government generally retains ownership of inventions developed on government property by its employees. However, employees may acquire rights in certain cases. For example, the Stevenson-Wydler Act, as amended by the Federal Technology Transfer Act of 1986 (P.L. 99-502), seeks to encourage technology transfer from government laboratories to industry under licenses, employee spin-offs, and limited partnerships.

Until 1980, it was also the general practice of government to retain ownership of inventions developed as a result of federal support at other institutions. Under P.L. 96-517 and subsequent amendments under the Trademark Clarification Act of 1984 (P.L. 98-620), small businesses and nonprofit organizations, which include universities and colleges, and government laboratories managed by these entities, can now receive all rights, title, and interest to inventions resulting from government-sponsored projects. P.L. 98-620 also permits exclusive licensing to companies, regardless of size.

The grant of rights under this law is subject to a nonexclusive license to the government and to regulations governing the reporting, patenting, and licensing of inventions. Should the contractor or grantee institution decide not to exercise its rights, the rights must be released to the government, which may then pursue a patent or have the technology released to the public by publication. An inventor wishing to acquire ownership rights must submit a petition to the government.

Policies at universities and nonprofit corporations must provide for government rights in the event of federally sponsored inventions, and ownership rights are often a subject of negotiation in regard to industry-sponsored research at these institutions. Policies at these institutions generally also allow for release of an invention to the inventor if a decision is made not to pursue patenting or seek a licensee or if the institution fails to act relative thereto in a timely manner.

IMPLICATIONS FOR LICENSORS

The federal research program does not have as its objective the production of inventions for commercialization; neither is this the mission of the university and nonprofit community. Nevertheless, government, universities, and the nonprofit research community do recognize that inventions that might benefit the public are often produced in the course of research. It is towards furthering this public benefit, and achieving the resulting rewards that inventions policies have been instituted by these parties. While licensor institutions are the recipients of the research money, actually the process involves project-specific grants or contracts for individual investigators or research teams, who are usually employees of the licensor institution.

The inventions policy of the federal government is directed to the institutions that manage the grant and research contract instruments. It encourages grantee/contracting (licensor) institutions to actively seek licenses, ensures technology development incentives for investigators, and requires a share of the rewards of commercialization for research. The government's ability to carry out its policy comes primarily from its authority to regulate the practices of those who administer federal grants and contracts. These are the basic instruments under which individuals or licensor institutions receive federal money for research.

Grants and Contracts

Under the federal grant and contract system, federal awards made to investigators undergo a pre-award stage and a post-award stage. During the pre-award stage, the research proposal is developed and reviewed. The process strives toward competitiveness. Funding opportunities are made known to the public in advance, prompting proposal development and submission. This pre-award stage reflects the competitive aspects of government procurement policy. By its disclosure of research needs in widely read publications, the government ensures that all candidates become aware of funding opportunities. Usually, this includes ranges for the amounts available per project and the purposes to be accomplished. Selection of the best candidate is then sought by having a panel of known experts in the field conduct an evaluation of the proposals. Their recommendations serve as the basis for determining who get the awards. This evaluation procedure has come to be known as the peer-review system.

Science-based technological development depends on the continuing support of basic science, and the government's definition of research has a significant bearing on the kinds of proposals it receives. Which proposals get funded either pave the way to future inventions or contribute to the cumulative knowledge base upon which progress depends. Reviewers therefore, are challenged to determine responsible trade-offs. The best choice between present and future technological needs is achieved if the reviewing body is comprised of experts from both the noncommercial and commercial sectors.

If approved, the award enters a post-award process. It is received and administered by licensor institutions on behalf of the investigator(s), one of whom is designated as the project director, with invention rights accruing, as discussed above. The research might or might not result in an invention. Under the law, the institution is given two years to elect to retain title, and if it does so, an additional one year within which to file a patent application if the invention has not been publicly disclosed.

Because of the costliness of patenting, an important consideration for this decision is the market relevance of the technology and the presence of industrial interest. By requiring a patenting decision within a specified time frame, grantee institutions are required to act diligently to ensure licensing or technology transfer. Failure to comply results in forfeiture of rights to the invention.

While the patent policies of most institutions provide for royalty sharing with the inventor, this is also required under government policy. The provision is that part of the royalty should go to the inventor and part to the support of research, after allowances have been made for administration and patenting costs. Accordingly, rewards to the inventor stimulate inventorship, and the continuing support of research is fostered by the share dedicated to that purpose.

Conflict of Interest

A perennial concern on the part of both federal sponsors and grantee institutions is that the lure of financial rewards from inventions and associations with business entities may undermine objectivity and the instruction and basic-research missions of these institutions. Employees and/or faculty, realizing the potential for such payoffs, may conduct themselves, direct efforts over which they have influence, or use space over which they have control, in a manner that unduly favors organizations promising them personal gain.

Fear that publicly sponsored research may be compromised as a result has prompted government agencies to action. In June 1995, the National Science Foundation's Investigator Financial Disclosure Policy became effective. The Department of Health and Human Services (DHHS) has issued a Notice of Proposed Rule Making closely resembling the NSF policy for public comment, and other agencies are expected to follow suit. Both NSF and DHHS agree with critics that a uniform, government-wide policy would be preferred. However, the indications at this time are that the agencies will be acting separately.

The NSF puts the responsibility for collecting and reviewing investigator financial information and managing conflict on the licensor institution. Basically, the recipient institutions are required to maintain an appropriate written and enforced conflict-of-interest policy that ensures disclosure by the investigator of any financial interest owned by the investigator or family members that would be materially affected by the NSF-sponsored work. The disclosure must be made at the time of submission of the research proposal to the NSF.

The Institutional Conflict of Interest Policy required by NSF must further designate a person to review and resolve conflicts, must contain mechanisms of enforcement, and must specify arrangements to keep NSF informed. Records relative thereto must be maintained until the lesser of three years after expiration of the award or until resolution of government action concerning the records. However, the NSF requirements may be waived if the reviewer determines that their imposition will serve no useful purpose or if the negative effects of the policy are outweighed by the benefits of scientific progress, technology transfer, or public health.[4]

IMPLICATIONS FOR LICENSEES

Business is the prime agent for translating technology into new products and processes to benefit the public. Technology transfer enables a business to acquire the title to a new technology or to license the rights of ownership from those who own the title. The availability of inventions for licensing and the actions government takes to foster collaborations with industry are also important for the success of technology transfer.

Business Access to Ownership Rights

A company's desire to obtain the rights to a new technology depends in large measure on the significance of the new technology for strengthening the business, its potential to make profits for the business, and the ease with which the transfer of title is likely to occur. Important factors in a business's decision making process are the access it has to the full possibilities of the invention and the freedom it has to fully exploit the market potential of the technology. These factors figure significantly in the technology transfer process, and they are affected by title protections provided in patent and trade secret law.

Most employers require their employees to assign title to new inventions to them. Therefore, while inventors have a constitutional right to their inventions, employers have a right to request and acquire title as a condition of employment. The result is an inducement to technological progress, since those who have the means to sponsor research are able to invest in inventivity without fear or challenge to their ownership rights in the resulting products and processes.

Some companies reward individual employees for their inventivity. Other companies argue that monetary reward encourages self-interest and competitiveness with coworkers. The "teamwork" emphasis is more likely to produce one strong patent application, while encouragement of individual efforts is more likely to result in a number of minor patent applications.[5] This freedom that companies have under the law removes ownership rights as an obstacle in product and process innovation.

Business Access to Government-Sponsored Inventions

The licensees of inventions are companies that have adopted licensing as a business strategy. These companies might depend exclusively on outside sources for innovative ideas, or they might in varying degrees use external invention to complement their own in-house product and process development. The provisions of the technology transfer statutes, incentives arising from the procurement demands of mission-oriented research, and policies relating to dissemination of research results attempt to reach out to interested companies. Both regulatory and financial considerations apply.

Regulations of the Bayh-Dole Act apply with respect to the companies to be licensed, the location of business, and the rights of the federal government. It also allows small business to acquire title if the invention is developed on company facilities. From the viewpoint of the licensees, these regulations can have both beneficial and adverse effects.

The Bayh-Dole Act requires that small businesses be given first preference as licensees of inventions arising under federally sponsored research, and that such preference be for U.S. based companies, unless this is not feasible. For the small company, the arrangement alleviates the threat of having to compete with a large company for a license. As noted previously, licensors are more apt to favor a large company that has greater resources to develop and market the product extensively, thereby generating larger profits and royalties. However, the regulation introduces an extra element of risk for the large company.

For example, if a large company takes the license, and a small company later demonstrates that manufacture by the small company is feasible and that the licensor institution failed to exercise due diligence in seeking a small company, the government could take action under its March In Rights to require a transfer of the license to the small company. This possibility could deter large companies from licensing federally sponsored inventions.

Furthermore, the Small Business Innovation Act of 1982 (P.L. 97-219) requires agencies to fund small business R&D that is related to the agency's mission. This support is administered by the Small Business Innovation Research (SBIR). In subsequent legislation, the Small Business Technology Transfer (STTR) Program of 1992 (P.L. 102-564), five major agencies are charged with sponsoring cooperative R&D projects involving a small company and researchers in noncommercial entities.[6]

Business Access to Government-Owned Inventions

The Stevenson-Wydler Act applies to inventions that are directly owned by the federal government. It seeks to develop a public infrastructure to foster private/public collaboration in market-relevant research and development and in mechanisms for the transfer of federally owned inventions.[7] The Act also mandates that agencies allocate 0.5 percent of their R&D budget for transfer activities. Establishment of industrial extension services occurs under the Omnibus Trade and Competitiveness Act of 1988 (P.L. 100-418), and Executive Order 12591 promotes cooperative R&D efforts among the various levels of government.

Federally funded centers, offices, and services therefore, promote technology development and transfer. The National Technical Information Service, founded under the Stevenson-Wydler Act, is a coordinator and information clearinghouse for federally owned or originated technologies. The National Technology Transfer Center (NTTC) conducts training programs in collaboration with other agencies and industry to advance awareness, knowledge, and commercial applications of these technologies.

The federal Centers are sometimes operated specifically within one agency and/or specific regions. For example, NASA has six Regional Technology Transfer Centers. An example of collaboration is the agreement between NTTC and the Center for Technology Commercialization (CTC) to share information, study how their services are being used, and discuss how they might be improved. CTC serves U.S. companies located in the Northeast; its mission is to open the doors to technologies developed by NASA and other federal agencies. Another partnership is between NTTC and Knowledge Data Express Systems, a private entity, directed toward the dissemination of information on federal, university, and industry capabilities.[8]

The Federal Technology Transfer Act also provides for what has become known as the "CRADA's," or Cooperative Research and Development Agreements, between government-owned and operated laboratories and the private sector. The requirements of the Bayh-Dole Act apply to the CRADA's, and they make it easier for directors of laboratories to acquire ownership rights to inventions. In 1989, this agreement became available to government-owned, contractor-operated laboratories, under the National Competitiveness Technology Transfer Act (P.L. 101-189).[9]

Federal Research and Competition

Companies are continually pressed to higher levels of technological achievement in order to remain competitive as providers in the nation's mission-oriented research program. They are particularly affected by the needs of national defense. R&D for strategic defense is characterized by a reliance on industry and is concentrated in a few product categories. The products receiving the greatest direct benefit are aircraft and missiles and electrical communications equipment, which accounted for 78.6 percent of the federal funds received by industry in 1990.[10]

Federally sponsored industrial research support is also concentrated in very large companies, 87.6 percent of the total going to companies with more than 25,000 employees. Companies with fewer than 5,000 employees received 6 percent. The amount received by the smaller companies is widely diffused and accounts for a significant portion of their individual research budgets. Smaller companies also benefit from the procurement requirement of "second sourcing."[11] The consequence of the funding patterns and secondary effect of government procurement practices is a widespread challenge to companies to excel technologically, in order to remain competitive.

Antitrust Policy

While intellectual property has certain distinguishing characteristics, it also possesses attributes that subject it to the same antitrust principles as apply to other forms of tangible and intangible property. For example, the right to exclude others is enjoyed by owners of other private property. However, the market power conferred by the monopoly rights arising from the patent, as with any supracompetitive business undertaking, is not in itself unlawful. In general, it is the conduct arising out of the possession of such power that gives rise to illegalities.

The case for antitrust law in technology policy is based on the belief that technological progress is inhibited if firms are not sufficiently competitive. In technology policy, this refers primarily to the exercise of intellectual property rights. Because this law gives rise to punishable illegalities in patenting and licensing behavior, firms will usually avoid behavior in those areas that may be construed as violations of the antitrust law.

The antitrust statutes most relevant for patents are the Sherman Act and the Clayton Act. Under Section 1 of the

Sherman Act, it is unlawful to combine or conspire in restraint of trade; Section 2 prohibits monopolies and conspiracies as well as attempts to monopolize. Section 3 of the Clayton Act proscribes business practices in sales, leasing, pricing, and other areas that would have the effect of substantially lessening competition or that would tend to create a monopoly.

Determinations by the Justice Department of violations of the law occur under either the rule of reason or the per se doctrine. The rule of reason involves a detailed inquiry into the purpose and effect of the restraint of trade under investigation. Under the per se doctrine, such acts can be declared unlawful without any supporting inquiry, due to the blatant obviousness of the offense. These acts include "naked price-fixing, output restraint, and market division among horizontal competitors, as well as certain group boycotts and resale price maintenance."[12]

In 1994, the Justice Department issued draft Intellectual Property Guidelines, covering the licensing and acquisition of intellectual property protected by patents, trade secret law, and copyright. In an accompanying press release, a member of the Antitrust Division task force maintained that the Guidelines would ensure sound antitrust enforcement. The intention of the Guidelines is to reduce uncertainties in the law; since fear of breaking antitrust law can be an inhibiting factor in technological development.

Licensing is viewed as procompetitive conduct. However, when licensing interrupts competition that would otherwise occur in the absence of the license, the act is suspect. This may happen when licensing involves actual or potential competitors. The Guidelines seek to protect competition in the goods, technology (substitute goods), and innovation markets.

Innovation markets are a new element. In this area, the Justice Department would challenge the act if the parties involved are also among the few whose research and development will produce innovations. In these cases, the competition in research and development will also be taken into account in assessing anticompetitive effects.

The anticompetitiveness of the transaction is determined by the degree of market control collectively available to the parties. In this regard, the Guidelines have designated a "safety zone." Accordingly, parties coming together in a license arrangement face little risk of being challenged by the Justice Department if their combined share of the market for the product licensed does not exceed 20 percent. If an anticompetitive effect is determined to be the case, the Department will then inquire as to whether the procompetitive benefits arising therefrom outweigh the anticompetitive effects.

Generally, the rule of reason is invoked in the case of restraints in intellectual property licensing. To rule out application of the per se rule, a determination must be made as to whether the suspect license arrangement contributes to an efficiency-producing integration of economic activity. This concept of efficiencies refers to benefits accruing as a result of enhancing the development and commercialization of the licensed product or of reducing the transactions costs brought about by the affiliation of the parties. If there is an efficiency-producing integration of economic activity, the rule of reason will be applied.

Arrangements that give rise to challenges from the Justice Department can occur under horizontal restraints, resale price maintenance, tying arrangements, exclusive dealings, cross-licensing and pooling arrangements, and grantbacks. Therefore, anticompetitive effects may be construed if the arrangement involves restrictions on the licensee's resale price of the licensed product, is conditional upon purchase by the licensee of another product that is unnecessary for the practice of the licensed technology, restricts purchases by the licensee of competitor technologies, puts competitors at a disadvantage, and/or retards innovation by discouraging research and development.

The Guidelines, by reducing the uncertainty in antitrust law, will alleviate inaction among small businesses and innovators due to fear of the unknown. It is expected that this will be beneficial for technology transfer. However, sound antitrust enforcement is also expected, which may cause large firms to refrain from supplementing in-house research and development with externally developed opportunities, due to the increased likelihood of challenges from the Justice Department.[13] However, pre-competitive collaborative research with other large companies is encouraged.

Corporate interest in cooperation on pooled research is allowed under the Cooperative Research Act of 1984 (P.L. 98-462). This Act frees firms wishing to combine their resources for purposes of pre-competitive R&D, from treble damage concerns. The Act has already resulted in the formation of major consortia, such as the Semiconductor Research Corporation (SRC) and the Microelectronics and Computer Technology Corporation (MCC).[14]

INDICATORS OF TECHNOLOGICAL PROGRESS

According to *The Process of Economic Growth* by W. W. Rostow, the United States reached its technological maturity at the turn of the century. Rostow defines technological maturity as a period when a country has effectively applied the range of (then) modern technology to the bulk of its resources. During the 20th century, the United States has advanced beyond technological maturity to an age of high mass consumption.

The age of high mass consumption is one stage before Rostow's final stage, which is that of an economy advancing beyond mass consumption. Entry into high mass consumption occurred in the 1920s, during which time the economy offered, ". . . enlarged private consumption—including single family homes, and durable goods and services—on a mass basis."[15]

Technological progress is an international phenomenon, proceeding at different rates in the different countries of the world. Nations have access to a common technological literature, and provide each other with motives for advancement. In this process, the United States holds a dominant position in science and technology, when viewed in terms of the scale of research and development conducted, the education and performance of scientists and engineers, and the production of scientific publications, inventions, patents, and innovations.

For international trade comparisons, the National Science Board Report uses a broader definition of technology transfer. Its concept reflects demand at the international level for the high technology capability of U.S. business, as well as U.S. competitiveness, and includes the export of "technology-embodying" products, the expansion or establishment of subsidiaries through foreign investment, and the transfer of "disembodied-technology" through the sale of patent licenses and blueprints.

The data show that significant proportions of the output of important high technology industries are going to foreign buyers. Among the many industries represented are the producers of professional and scientific instruments, office machines and computers, plastics and resins, and agricultural chemicals. Of particular note, more than half of electrical transmission and distribution equipment and of the output of engines and turbines, aircraft and parts are produced for export.

Foreign investment is concentrated in a few European countries and Canada and involves primarily the chemical and machinery industries. For "disembodied-technology," data is provided in the form of royalty and fee payments for "arms-length" transactions. These are transactions that occur between unaffiliated firms. The receipts, exceeding $700 million came predominantly from other technologically advanced nations, again indicative of this country's competitive advantage.

Today, the United States is the largest performer of R&D among the world's market economies. It has the largest R&D work force and leads in science degrees granted, publications, and patents. Its performance, reported in the 1985 "Science Indicators," a publication of the National Science Board is indicative. This publication shows that 5 countries taken together, account for approximately 80 percent of the R&D conducted by the 24 countries of OECD. The United States is among these 5 countries, and accounts for about half of the 80 percent portion.[16] These data are also indicative of the size of the U.S. relative most to other countries.

Notes

[1] Henry Ergas, "Does Technology Matter," in *Technology and Global Industry*. Ed. by Bruce R. Guile and Harvey Brooks (Washington, D.C.: National Academy Press, 1987), 12.

[2] National Science Foundation, *National Patterns of R&D Resources* (1992), 76-78.

[3] U.S. Patent and Trademark Office, *Patenting by Organization* (February 7, 1994), 1.

[4] Investigator Financial Reporting. *Federal Register*, Vol. 59, No. 123 (June 28, 1994): 33308-33311.

[5] David E. Noble, *America by Design—Science, Technology and the Rise of Capitalism* (New York: Alfred A. Knopf, 1977), 101.

[6] U.S. Government, *Technology Transfer Legislative History*, in TechTRANSIT Home Page. Internet, http://www.dtic.dla.mil/... Legislative_History.html. (December 7, 1996), 1-4.

[7] Alan H. Einhorn, *Overview of Federal Law on Technology Transfer* (Manuscript distributed at 1994 Conference on Commercializing Biomedical Technologies), 12.

[8] Technology Touchstone, *NTTC/Knowledge Express Enter Partnership*, Vol. 3, No. 1 (March 1995): 1-4.

[9] Einhorn, *Overview*, 13.

[10] National Science Foundation, *National Patterns*, 76-78.

[11] Ergas, *Does Technology Matter*, 199.

[12] U.S. Department of Justice, *Justice Department Release of Draft Intellectual Property Guidelines for Public Comment* (August 8, 1994), 11.

[13] Ibid., 1-25.

[14] U.S. Government, *Legislative*, 1.

[15] W. W. Rostow, *The Process of Economic Growth* (New York: W. W. Norton & Company, Inc., 1962), 318-324.

[16] National Science Board, *Science Indicators, The 1985 Report* (National Science Foundation, 1985), 4-13.

Appendices

Appendix A

Sample New Technolgoy Evaluation Agreements

Appendix A - 1

THE CONFIDENTIALITY AGREEMENT

Confidentiality Agreement
between
[Name of PROVIDER] and [Name of COMPANY]

THIS AGREEMENT is effective upon the date of the last signature herein, and is between _______________, having its principal office at _________________(mailing address: "PROVIDER") and ________________, having its principal office at ________________ (hereinafter called "COMPANY").

THE PARTIES AGREE AS FOLLOWS:

1. COMPANY, agrees to keep in strict confidence and not to use any information, knowledge, data and/or know-how either contained in or related to "_____," (collectively called "INFORMATION"), for its research or commercial use (except for technical and economic evaluation internal to said COMPANY) without prior written consent acquired from

PROVIDER. It is further agreed that COMPANY, shall keep in confidence and not disclose any part of INFORMATION to a third party or parties for a period of five (5) years from the date hereof.

2. The period during which INFORMATION can be transferred under this Agreement shall extend six (6) months from date hereof, and this period may be earlier terminated by either party upon thirty (30) days notice to the other and may be extended from time to time by mutual agreement between the parties in writing.
3. Any obligation of COMPANY, as set forth in the preceding paragraphs shall not apply to any information, knowledge, data and/or know-how which:
   (a) is or hereafter becomes generally available to the public through no fault of COMPANY.
   (b) COMPANY can show was in its possession at the time of disclosure and was not acquired, directly or indirectly, from PROVIDER.
   (c) COMPANY can show was received by it from others that shall not have received the same, directly or indirectly, (d) COMPANY can show was independently developed by employees of COMPANY who have not had access to or knowledge of INFORMATION disclosed hereunder.
4. COMPANY agrees to obligate its employees that shall have access to any portion of INFORMATION to protect the confidentiality of the INFORMATION.

IN WITNESS WHEREOF, the parties hereto have respectively caused this agreement to be signed by their duly authorized representatives on the day set forth herein below.

COMPANY		PROVIDER	
BY:	______________	BY:	______________
TITLE:	______________	TITLE:	______________
DATE:	______________	DATE:	______________

Appendix A - 2

THE TESTING AGREEMENT

Testing Agreement
between
[Name of PROVIDER] and [Name of RECIPIENT]

THIS AGREEMENT, made and entered into this ________________ day of ________________, 1996, by and between ________________, a corporation duly organized and existing under the laws of the State of __________ and having a principal office (mailing address: P.O. Box 9, Albany, New York 12201-0009) (hereinafter referred to as the "PROVIDER"), and ________________, a corporation duly organized under the laws of the State of __________, with its principal offices at ________________ (hereinafter referred to as "RECIPIENT").

WITNESSETH

WHEREAS, the PROVIDER is the owner of certain proprietary compounds and information developed by ________________ and has the right to grant licenses under said proprietary material and information; and

WHEREAS, RECIPIENT desires to obtain a sample of such compounds solely for testing purposes to determine if activity levels meet market requirements, upon the terms and conditions hereinafter set forth.

NOW, THEREFORE, in consideration of the premises and the mutual covenants contained herein, the parties hereto agree as follows:

ARTICLE 1
OBLIGATIONS OF THE PARTIES

1.1 "COMPOUND" shall mean ______________.

1.2. Neither the COMPOUND supplied, nor other materials derived in whole or in part from the original COMPOUND, may be transferred to any third party, or disclosed to the public without the PROVIDER's written authorization, since such COMPOUND remains the property of the PROVIDER. It is further agreed that RECIPIENT, shall keep in confidence and not disclose any related information to a third party or parties for a period of five (5) years from the date hereof.

1.3. Any obligation of COMPANY, as set forth in the preceding paragraphs shall not apply to any information, knowledge, data or know-how ("INFORMATION") which:

(a) is or hereafter becomes generally available to the public through no fault of COMPANY.

(b) COMPANY can show was in its possession at the time of disclosure and was not acquired, directly or indirectly, from PROVIDER.

(c) COMPANY can show was received by it from others that shall not have received the same, directly or indirectly, from PROVIDER.

(d) COMPANY can show was independently developed by employees of COMPANY who have not had access to or knowledge of INFORMATION disclosed hereunder.

(e) COMPANY agrees to obligate its employees that shall have access to any portion of INFORMATION to protect the confidentiality of the INFORMATION.

1.4 This Agreement shall be construed, governed, interpreted and applied in accordance with the laws of the State ________________.

1.5 For the rights and privileges granted under this Agreement, RECIPIENT will pay to the LICENSOR the amount of __________ thousand dollars ($__________), upon receipt of the COMPOUND.

1.6 This Agreement shall be for a period of one year, beginning from the date of delivery of COMPOUND to RECIPIENT. Upon expiration of the Agreement, RECIPIENT shall promptly return COMPOUND to PROVIDER.

1.7 During the term of this Agreement and for a period of three (3) months thereafter, RECIPIENT shall have an option to negotiate rights to an exclusive license to manufacture and sell products incorporating the COMPOUND. During the period of this option PROVIDER shall refrain from negotiating terms of a license agreement with other companies and shall not grant a license or licensing rights to any other company.

ARTICLE 2
HOLD HARMLESS AND WARRANTY

2.1 The RECIPIENT shall be responsible for all damages to life and property which arise from use of the COMPOUND which is supplied hereunder. The RECIPIENT agrees to indemnify and hold harmless the PROVIDER, against any and all claims, damages, and expenses of whatsoever nature arising from, growing out of, or relating to the RECIPIENT's acceptance and use of the COMPOUND being provided hereunder. The RECIPIENT agrees to reimburse the PROVIDER for any damages which the RECIPIENT may incur as a result of the RECIPIENT's acceptance and use of the COMPOUND.

2.2 The PROVIDER makes no representations whatsoever as to the COMPOUND, which are provided hereunder without warranty of merchantability or fitness for a particular purpose or any other warranty, express or implied, including, but not limited to warranties or representations as to the purity, activity, safety, or usefulness of the COMPOUND.

IN WITNESS WHEREOF, the parties hereto have set their hands and seals and duly executed this agreement the day and year first above written.

[Name of PROVIDER]	[Name of RECIPIENT]
______________________	______________________
Title: _______________	Title: _______________
Date: _______________	Date: _______________

Appendix A - 3

THE SCREENING AGREEMENT

Screening Agreement
between
[Name of PROVIDER] and [Name of COMPANY]

THIS AGREEMENT, effective as of the _____ day of _____, _____ between [name of provider] existing under the laws of ______, and having its principle office at _________________ (mailing address _________________), (hereinafter referred to as "PROVIDER")and [name of company], having its principle address at _______________ (hereinafter referred to as "COMPANY").

WITNESSETH

WHEREAS, PROVIDER has certain compounds which provider wishes to have subjected to a program of screening and evaluation for the purpose of determining commercial utility; and

WHEREAS COMPANY is interested in obtaining compounds from PROVIDER for the purpose of screening and evaluation for use in the Field as defined hereinafter, referred to below as the "PROGRAM."

NOW, THEREFORE, in consideration of the premises and the mutual covenants and agreements contained herein it is mutually agreed by and between the respective parties as set forth below.

ARTICLE 1
DEFINITIONS

1.1 "FIELD" means chemical compounds relating to (define area of interest).

1.2 "PROGRAM" means the screening and evaluation of chemical compounds in the FIELD.

ARTICLE 2
COMPOUND SCREENING AND EVALUATION

2.1 At the PROVIDER's option PROVIDER will provide to COMPANY a list of compounds synthesized by PROVIDER that are available for the PROGRAM.

2.2 Upon receipt of the list of compounds, COMPANY shall select for the PROGRAM those compounds COMPANY believes may have activity in the FIELD taking into account the compound structure and pertinent information already known to COMPANY. Any compound not selected shall not be subject to this Agreement. For each compound selected by COMPANY, COMPANY shall pay to PROVIDER the sum of $_____ per gram of the compound provided to COMPANY by PROVIDER.

2.3 If after completion of its evaluation of a particular compound COMPANY determines that it has no further interest in that compound, COMPANY will provide PROVIDER with a summary of the primary screening results for such compound. COMPANY will complete its evaluation within 9 months. The summary of the screening results may be published by PROVIDER in any manner deemed desirable. However, COMPANY may be given the opportunity to review, prior to publication, any proposed publication containing such summary or any portion thereof. Such review would be only for COMPANY's information, and in no way connotes a restriction on PROVIDER's right to disseminate such publication.

2.4 If, after completion of its evaluation of a particular compound, COMPANY determines that it has further interest in that compound, COMPANY will discuss with PROVIDER mutually suitable arrangements for further testing and possible synthesis of the compounds and analogues thereof. With respect to each compound in which COMPANY has further interest, PROVIDER will provide COMPANY with the following available information possessed by PROVIDER:
   (a) whether or not the compound is considered to be new;
   (b) whether or not the compound has been screened for biological activity and, if so, a summary of screening results;
   (c) whether the compound was synthesized with under support of another entity;
   (d) whether or not any other party has any rights to such compound;
   (e) information on closely related compounds; and
   (f) any technical information related to the compound and analogues thereof.

ARTICLE 3
CONFIDENTIALITY

3.1 COMPANY agrees to hold in confidence all information related to compounds selected for screening which is received from PROVIDER under this Agreement for five (5) years from the date of receipt of such information without the express written permission from PROVIDER, provided such information is in writing and is marked "Confidential." Further, COMPANY will not be obligated to hold in confidence any such information that COMPANY can show in writing:
   (a) was in COMPANY's possession as of the date of its receipt hereunder as evidenced by our written records; or

(b) was available to the public as of the date of its receipt hereunder; or
(c) is published or otherwise becomes available to the public following its receipt hereunder not as a result of any action by COMPANY; or
(d) is disclosed to COMPANY on a non-confidential basis by a third party who has a right to make such disclosure; or
(e) as can be evidenced by written records is developed for COMPANY by persons who have not had direct or indirect knowledge of compounds and; or information obtained from the PROVIDER.

3.2 COMPANY shall limit access of compounds and information obtained from PROVIDER to only those persons within COMPANY that have a need to know for the PROGRAM and agrees to obligate such employees to protect the confidential and propriety nature of the compounds and related information.

ARTICLE 4
PATENTS AND LICENSING RIGHTS

4.1 All patents arising out of the PROGRAM shall belong to PROVIDER or to COMPANY or to both, in accordance with the laws of the United. Prior to COMPANY deciding to proceed to field test, develop or to commercialize any product resulting from the PROGRAM, COMPANY agrees to negotiate a mutually satisfactory royalty bearing license agreement with PROVIDER under any patent rights PROVIDER has or may have in any country necessary for COMPANY to make, use or sell the product in the FIELD. COMPANY shall make its interest in a license known in writing to PROVIDER within six (6) months after completion of its evaluation as provided in Paragraph 2.3 above. During this time the parties may either negotiate an extension of the evaluation period, an option to a license or a license agreement. Unless continued by mutual consent of the parties any rights of COMPANY to a particular compound selected for evaluation shall expire at the end of this six (6) month period.

ARTICLE 5
HOLD HARMLESS

COMPANY agrees to indemnify PROVIDER and hold it harmless from any action, claim or liability without limitation, liability for death, personal injury, or property damage, arising directly or indirectly from COMPANY's possession, testing, screening distribution or other use of compound provided by PROVIDER under this agreement, and/or from COMPANY publication or distribution of test reports, data and other information relating to said compounds, except, however, if such action, claim or liability is directly and principally caused by or is the result of negligence or the intentional acts of PROVIDER.

ARTICLE 6
NOTICES

Any notice or report required or permitted to be given under this Agreement shall be deemed to have been sufficiently given for all purposes if mailed by first class registered mail to the following addresses of either party:

TO PROVIDER: ______________________________

____

____

TO COMPANY: ______________________________

____

____

or to such other address as shall hereafter have been furnished by written notice by such party to the other party. Notices shall be deemed given as of the date mailed.

ARTICLE 7
TERM

The term of this PROGRAM shall be for three (3) years from the effective date. Provided however, that either party may terminate the PROGRAM at will upon three (3) months prior written notice to the other party. Termination of the PROGRAM shall not terminate the obligations imposed on the parties under Paragraphs 2.3 and 2.4, and Articles 3 and 4.

ARTICLE 8
MISCELLANEOUS

8.1 This Agreement is not assignable in whole or in part, and the Agreement shall be binding up on and inure to the benefit of the respective successors of PROVIDER and COMPANY.
8.2 The parties agree not to disclose the nature of this Agreement nor use the name of the other in any promotional manner without prior written permission of the other.
8.3 This Agreement shall be construed as having been made under the laws of the State of __________.
8.4 With regard to COMPOUNDS submitted and screened, COMPANY will undertake to comply with applicable requirements of a sponsoring entity, if any, under which the compounds were synthesized.
8.5 This Agreement represents the entire understanding between the parties and supersedes any and all previous understanding either oral or written with respect to the

subject matter and may not be amended, supplemented or otherwise modified except by an instrument in writing signed by both parties.

8.6 IN WITNESS WHEREOF, the parties hereto have set their hands and seals and duly executed this agreement the day and year first above written.

[Name of PROVIDER]		[Name of COMPANY]	
By:	________________	By:	________________
Title:	________________	Title:	________________
Date:	________________	Date:	________________

Appendix A - 4

THE OPTION AGREEMENT

Option Agreement
between
[Name of PROVIDER] and [Name of COMPANY]

THIS AGREEMENT, effective as of the _____ day of ________, _____ between [name of provider] existing under the laws of _____, and having its pri¨nciple office at _________________, (mailing address _________________), (hereinafter referred to as "PROVIDER")and [name of company], having its principle address at _________________ (hereinafter referred to as "COMPANY").

WITNESSETH

WHEREAS, PROVIDER is the owner of certain proprietary information which may or may not be patentable relating to an invention in the field of __________ (hereinafter defined); and

WHEREAS, PROVIDER is interested in disclosing the invention to COMPANY for the purpose of determining its commercial utility; and

WHEREAS, COMPANY is interested in obtaining information relating to the invention for the purpose of determining its interest in licensing.

NOW, THEREFORE, in consideration of the premise and the mutual covenants and agreements contained herein it is mutually agreed by and between the respective parties as set forth below.

ARTICLE 1
DEFINITIONS AND OBLIGATIONS
OF THE PARTIES

1.1 "TECHNOLOGY" means (define the invention).
1.2 PROVIDER shall provide to COMPANY a complete description of the TECHNOLOGY within thirty (30) days after the execution of this Agreement.
1.3 Upon receipt of the information describing the TECHNOLOGY, COMPANY shall pay to the PROVIDER an option fee in the amount of _____ dollars ($_____).
1.4 The term of this Agreement shall be for one (1) year from the first date written above.
1.5 During the term of this Agreement, COMPANY shall have an option to negotiate rights to an exclusive license to manufacture and sell products incorporating the TECHNOLOGY. During the period of this option PROVIDER shall refrain from negotiating terms of a license agreement with other companies and shall not grant a license or licensing rights to any other company.
1.6 If after completion of its evaluation of the TECHNOLOGY COMPANY determines that it has no further interest in the TECHNOLOGY, COMPANY will provide to PROVIDER an explanation of the technical and market limitations of the TECHNOLOGY.

ARTICLE 2
CONFIDENTIALITY

2.1 COMPANY agrees to hold in confidence all information related to the TECHNOLOGY received from PROVIDER under this Agreement for five (5) years from the date of receipt of such information and shall not disclose such without the express written permission from PROVIDER, provided such information is in writing and is marked "Confidential." Further, COMPANY will not be obligated to hold in confidence any such information that COMPANY can show in writing:

(a) was in COMPANY's possession as of the date of its receipt hereunder as evidenced by our written records; or
(b) was available to the public as of the date of its receipt hereunder; or
(c) is published or otherwise becomes available to the public following its receipt hereunder not as a result of any action by COMPANY; or
(d) is disclosed to COMPANY on a non-confidential basis by a third party who has a right to make such disclosure; or
(e) as can be evidenced by written records is developed for COMPANY by persons who have not had direct or indirect knowledge of the TECHNOLOGY and; or information obtained from the PROVIDER.

2.2 COMPANY shall limit access to TECHNOLOGY and information obtained from PROVIDER to only those persons within COMPANY that have a need to know and agrees to obligate such employees to protect the confidential and proprietary nature of the TECHNOLOGY and related information.

ARTICLE 3
HOLD HARMLESS

COMPANY agrees to indemnify FOUNDATION and hold it harmless from any action, claim or liability without limitation, liability for death, personal injury, or property damage, arising directly or indirectly from COMPANY's possession, testing, or other use of TECHNOLOGY provided by PROVIDER under this agreement, and/or from COMPANY publication or distribution of test reports, data and other information relating to said TECHNOLOGY, except, however, if such action, claim or liability is directly and principally caused by or is the result of negligence or the intentional acts of PROVIDER.

ARTICLE 4
NOTICES

Any notice or report required or permitted to be given under this Agreement shall be deemed to have been sufficiently given for all purposes if mailed by first class registered mail to the following addresses of either party:

TO PROVIDER:	TO COMPANY:
______________________	______________________
______________________	______________________
______________________	______________________

or to such other address as shall hereafter have been furnished by written notice by such party to the other party. Notices shall be deemed given as of the date mailed.

IN WITNESS WHEREOF, the parties hereto have set their hands and seals and duly executed this agreement the day and year first above written.

[Name of PROVIDER]		[Name of COMPANY]	
By:	________________	By:	________________
Title:	________________	Title:	________________
Date:	________________	Date:	________________

Appendix B

Royalty Rate Negotiations Spreadsheets

Appendix B - 1

Calculations Using Dollar Amounts Only

Variables	Amount		Royalty Impact	
A	B(1)	B(2)	C(1)	C(2)
Var. 1	50,000	60,000	3.55	4.26
Var. 2	30,000	30,000	2.25	2.25
Var. 3	(10,000)	(10,000)	(0.70)	(0.70)
Var. 4	(20,000)	(20,000)	(1.45)	(1.45)
Var. 5	20,000	20,000	1.45	1.45
Totals	70,000	80,000	5.00	5.71

Column A: The variables deemend to impact royalties.

Column B(1): The dollar amount assigned to each variable.

Column B(2): The new dollar amounts after the first round of negotiations. Note, only Var. 1 has been changed, from $50,000 to $60,000.

Column C(1): The distribution of a beginning 5% royalty based on the ratios in Column B; e.g., 50,000/70,000 x 5% = 3.55 for Var. 1.

Column C(2): The ratio of each element in B(2) to its counterpart in B(1) multiplied by the corresponding element in C(1).

The total of Column C(2) is the new royalty rate, reflecting the impact of an increase of $10,000 in Var. 1.

Appendix B - 2

Calculations Using Importance Weights Only

Variables	Weight		Royalty Impact	
A	B(1)	B(2)	C(1)	C(2)
Var. 1	100	100	1.88	1.88
Var. 2	75	75	1.40	1.40
Var. 3	(50)	(50)	(0.94)	(0.94)
Var. 4	60	80	1.13	1.51
Var. 5	(25)	(25)	(0.47)	(0.47)
Totals	160	180	3.00	3.38

Column A: The variables deemed to impact royalties.

Column B(1): The weight assigned to each variable.

Column B(2): The new weights after the first round of negotiations. Note, only Var. 4 has been changed, from 60 to 80.

Column C(1): The distribution of a beginning 3% royalty rate based on the ratios of each element in Column B(1) to the total for B(1).

Column C(2): The division of each element in Column B(2) by its counterpart element in Column B(1), multiplied by the corresponding element in Column C(1).

The total of Column C(2) is the new royalty rate, reflecting the impact of an increase of from 60 to 80 in Var. 4.

Appendix B - 3

Calculations Including Both Dollars and Weights

Variables	Amount		Weight		Amount x Weight	
A	B(1)	B(1)	C(1)	C(2)	D(1)	D(2)
Var. 1	5,000	5,000	100	80	500,000	400,000
Var. 2	3,000	5,000	75	75	225,000	375,000
Var. 3	(1,000)	(1,000)	60	60	(60,000)	(60,000)
Var. 4	2,000	2,000	25	25	50,000	50,000
Var. 5	(4,000)	(4,000)	30	30	(120,000)	(120,000)
Totals					595,000	645,000

(Continued from above)

	Royalty Impact E(1)	E(2)
Var. 1	4.20	3.36
Var. 2	1.89	3.15
Var. 3	(0.50)	(0.50)
Var. 4	0.42	0.42
Var. 5	(1.01)	(1.01)
Totals	5.00	5.42

Column A: The variables deemed to impact royalties.

Column B(1): The dollar amount assigned to each variable.

Column B(2): The new dollar amounts after the first round of negotiations. Note, Var. 2 has been changed, from $3,000 to $5,000.

Column C(1): The Weight assigned to each variable.

<u>Column C(2)</u>: The new weights after the first round of negotiations. Note, the Weight for Variable has been dropped from 100 to 80.

<u>Column D(1)</u>: The elements in Column B(1) multiplied by their counterparts in Column C(1).

<u>Column D(2)</u>: The elements in Column B(2) multiplied by their counterparts in Column C(2).

<u>Column E(1)</u>: The Total of this column is given as the starting royalty rate (5%), and its parts are obtained by multiplying 5% by the ratios for each elements in Column D(1) to the total for Column D(1); e.g., the impact for Variable 1 is 500,000/595,000 x 5% = 4.20.

<u>Column E(2)</u>: This column shows the impact on the Royalty Rate of a change in the dollar value or the weight assigned to the variables. Each element in Column E(1) is multiplied by the ratio of the corresponding element Column D(2) to Column D(1); e.g., for the change in Variable 1: 400,000/500,000 x 4.20 = 3.36. The ratio is "1" for unchanged variables, therefore for these, the Column (h) elements are repeated in Column (i).

The 20 point decrease in the weight for Variable 1 decreased the royalty rate by .84 percentage points (4.20 - 3.36). This was offset by a $2,000 increase in the value of Variable 2, which raised royalty by 1.26 percentage points (3.15 - 1.89), resulting in a net change in the royalty rate of .42 percentage points, from 5% to 5.42%.

Appendix C

Sample License Agreements

Appendix C - 1

THE "SHRINK-WRAP" LICENSE AGREEMENT

License Agreement

(to accompany sales)

1. By making use of the software, which is delivered with this LICENSE AGREEMENT, the purchaser of the software, also known as LICENSE, agrees to comply with all the terms and conditions of the LICENSE AGREEMENT.

2. In consideration of payment of the license fee for the software titled "________________" by the LICENSEE, ________________ hereafter called LICENSOR grants a non-exclusive, non-transferable license to the LICENSEE upon the terms and conditions set out below.

3. The LICENSEE agrees that it will use this software solely for its internal purposes and shall not copy, distribute or transfer to any persons other than for use by employees or students of the LICENSEE. Subject to these restrictions, the LICENSEE may make one copy of the software solely for back-up purposes.

4. All title, interest, rights and copyrights to this software and derivative products shall at all times remain the property of the LICENSOR, and the LICENSEE agrees to preserve LICENSOR's property rights.

5. Nothing in this Agreement shall be construed as conferring rights to use in advertising, publicity, or otherwise, the name of LICENSOR.

6. LICENSEE acknowledges that the software is being supplied with documentation "as is" without any accompanying services from LICENSOR and that any such services and payments as may be required for modifications and enhancements shall be separately negotiated.

7. This Agreement shall be governed by the laws of the State of _____.

Appendix C - 2

THE "INTERNAL USE ONLY" LICENSE AGREEMENT

License Agreement
between
[Name of LICENSOR] and [Name of LICENSEE]

THIS AGREEMENT, made and entered into this _____ day of __________, 1996, by and between _______________, a corporation duly organized and existing under the laws of the State of ___________ and having a principal office _______________ (hereinafter referred to as the "LICENSOR"), and _______________, a corporation duly organized under the laws of the State of __________, with its principal offices at _______________ (hereinafter referred to as "LICENSEE").

WITNESSETH

WHEREAS, the LICENSOR is the owner of certain proprietary material and information developed by _________________ and has the right to grant licenses under said proprietary material and information; and

WHEREAS, LICENSEE desires to obtain a license for its internal research purposes upon the terms and conditions hereinafter set forth.

NOW, THEREFORE, in consideration of the premises and the mutual covenants contained herein, the parties hereto agree as follows:

ARTICLE 1
OBLIGATIONS OF THE PARTIES

1.1 "CELL" shall mean cell which is needed to produce Monoclonal Antibody _____.

1.2 "PRODUCT" shall mean the purified and unpurified monoclonal antibody produced by the CELL, specific to _______________. _______________.

1.3 "MATERIAL" shall mean CELL and PRODUCT.

1.4 The LICENSOR hereby grants to LICENSEE, and LICENSEE accepts a ten-year, non-exclusive license to use the CELL to produce the PRODUCT solely for LICENSEE's internal research purposes.

1.5 Neither the MATERIAL supplied, nor other materials derived in whole or in part from the original MATERIAL, nor information relating thereto may be disclosed or transferred to any third party, including public or private culture depositories, without the express written permission of the LICENSOR, since such materials remain the property of the LICENSOR.

1.6 This Agreement shall be construed, governed, interpreted and applied in accordance with the laws of the State _________.

1.7 For the rights and privileges granted under this license, LICENSEE will pay to the LICENSOR a license fee in the amount of _________ thousand dollars ($_________).

ARTICLE 2
HOLD HARMLESS AND WARRANTY

2.1 The LICENSEE shall be responsible for all damages to life and property which arise from use of the MATERIALS which are supplied to the LICENSEE pursuant to this Agreement. The LICENSEE agrees to indemnify and hold harmless the LICENSOR, against any and all claims, damages, and expenses of whatsoever nature arising from, growing out of, or relating to the LICENSEE's acceptance and use of MATERIAL being provided hereunder. The LICENSEE agrees to reimburse the LICENSOR for any damages which the LICENSOR may incur as a result of the LICENSEE's acceptance and use of the MATERIAL.

2.2 The LICENSOR makes no representations whatsoever as to the MATERIAL. They are experimental and are provided without warranty of merchantability or fitness for a particular purpose or any other warranty, express or implied, including, but not limited to warranties or representations as to the purity, activity, safety, or usefulness of the MATERIAL.

IN WITNESS WHEREOF, the parties hereto have set their hands and seals and duly executed this license agreement the day and year first above written.

[Name of LICENSOR]	[Name of LICENSEE]
______________________	______________________
Title: ________________	Title: ________________
Date: ________________	Date: ________________

Appendix C - 3

THE REGULAR LICENSE AGREEMENT

License Agreement
between

____

and

____

THIS AGREEMENT, is made and entered into this ______ day of ____________, 19____, by and between ______________ a corporation duly organized and existing under the laws of the State of __________ and having a principal office at ________________ (mailing address: ________________) (hereinafter referred to as "LICENSOR") and ________________ a corporation of the State of __________________ having its principal office at ________________ (mailing address: ________________), (hereinafter referred to as "LICENSEE").

WITNESSETH

WHEREAS, LICENSOR is the owner of all rights, title and interest to any technology or discovery, whether patentable or not, made or conceived in performance of a research program under the direction of ______________, and involving the development of ______________; and

WHEREAS, LICENSOR has applied for patent protection on said ______________ in the ______________; and

WHEREAS, LICENSOR has the right to grant licenses to these technologies or discoveries so that they may be utilized in the public interest, and is willing to grant a license thereunder to LICENSEE so that said objective may be accomplished; and

WHEREAS, LICENSEE is desirous of obtaining certain rights and licenses from LICENSOR relating to the aforementioned technology and discovery (such technology and discovery hereinafter defined),

NOW, THEREFORE, in consideration of the premises and the mutual covenants contained herein, the parties hereto agree as follows:

ARTICLE 1
DEFINITIONS

For the purposes of this Agreement, the following words and phrases shall have the following meanings:

1.1 "LICENSEE" shall mean _______________ and any subsidiary of _______________.

1.2 "SUBSIDIARY" shall mean any corporation, company or other entity more than fifty percent (50%) of whose voting stock is owned or controlled directly or indirectly by LICENSEE.

1.3 "TECHNOLOGY" shall be limited to mean all inventions, discoveries, information, technical data or other know-how, whether patentable or not, related to the _______________ which LICENSOR has heretofore developed and is free to disclose and furnish to LICENSEE hereunder.

1.4 "LICENSED PRODUCT" shall mean any item, the manufacture, use or sale of which, whether as a single item or part of a kit, utilizes TECHNOLOGY.

1.5 "NET SALES" shall mean LICENSEE's billings for LICENSED PRODUCTS produced hereunder less the sum of the following:

(a) Discounts allowed in amounts customary in the trade for direct sales.

(b) Sales taxes, tariff duties and/or use taxes directly imposed and with reference to particular sales.
(c) Outbound transportation prepaid or allowed.
(d) Amounts allowed or credited on returns.

No deductions shall be made for commissions paid to individuals whether they be with independent sales agencies or regularly employed by LICENSEE and on its payroll, or cost of collections. LICENSED PRODUCTS shall be considered "SOLD" when payments are received.

1.6 "EFFECTIVE DATE" shall mean the first date written in this Agreement.

ARTICLE 2
GRANT

2.1 The LICENSOR hereby grants to LICENSEE a world-wide exclusive license to use and practice TECHNOLOGY to make, have made, use, lease, and/or sell LICENSED PRODUCTS, including the right to grant sublicenses to the full end of the term for which rights are granted unless sooner terminated as hereinafter provided.

2.2 In order to establish a period of exclusivity for LICENSEE, the LICENSOR hereby agrees that it will not grant any other license to make, have made, use, lease and/or sell TECHNOLOGY during the period of time commencing with the Effective Date of this Agreement and terminating with the expiration of the last to expire issued patent.

2.3 During the term of the Agreement, LICENSEE shall have the right to sublicense world-wide any of the rights, privileges, and licenses granted hereunder.

2.4 LICENSEE hereby agrees that every sublicensing agreement to which it is a party and which relates to the rights, privileges, and license granted hereunder shall contain a statement setting forth the date upon which exclusive rights, privileges and license hereunder shall terminate.

2.5 LICENSEE agrees that any sublicenses granted by it have privity of contract between LICENSOR and sublicensee such that the obligations of this Agreement are binding upon the sublicensee as if it were in the place of LICENSEE. LICENSEE further agrees to attach copies of Articles 2, 5, 7, 14, and 15 of this Agreement to all sublicense agreements.

2.6 LICENSEE agrees to forward to LICENSOR a copy of any proposed sublicense agreement for review and approval by LICENSOR. Such approval shall not be unreasonably withheld. LICENSEE further agrees to forward to LICENSOR annually a copy of all royalty reports received by LICENSEE from its sublicensees during the preceding twelve (12) month period.

2.7 The LICENSOR represents and warrants that it is the owner of TECHNOLOGY free and clear of all liens, claims and encumbrances and has the sole and unrestricted right to grant the license as set forth herein.

ARTICLE 3
DUE DILIGENCE

LICENSEE has represented to LICENSOR, to induce the LICENSOR to issue this license, that LICENSEE will use all reasonable speed to create and produce a commercially marketable product incorporating TECHNOLOGY. At one year intervals after execution LICENSEE shall demonstrate to LICENSOR that it has and is continuing to develop said product or products incorporating TECHNOLOGY in a diligent manner with all reasonable speed and continues to provide appropriate funding for development of said product or products. In the event that LICENSEE fails to do so or to continue to actively market the product during the term of this Agreement, LICENSOR shall have the right to terminate the Agreement pursuant to Article 6.2.

ARTICLE 4
ROYALTIES

For the rights and privileges granted under the license, LICENSEE will pay to LICENSOR in the manner shown below:

(a) a licensing fee in the amount of __________ dollars (U.S.), payable upon the signing of this agreement, and
(b) a royalty in an amount equal to __________ of the NET SALES of LICENSED PRODUCTS produced by LICENSEE, and
(c) a royalty equal to ___________ of all ___________ on LICENSED PRODUCTS received by LICENSEE from any sublicensee.

ARTICLE 5
REPORTS AND RECORDS

5.1 LICENSEE shall keep full, true and accurate books of account containing all particulars which may be necessary for the purpose of showing the amount payable to LICENSOR by way of royalty as aforesaid. Said books of accounts shall be kept at LICENSEE's principal place of business or the principal place of business of a Division of LICENSEE which is marketing the LICENSED PRODUCT. Said books and the supporting data shall be open at all reasonable times, for five (5) years following the end of the calendar year to which they pertain, to the inspection of LICENSOR's Internal Audit Division and/or of an independent certified public accountant retained by LICENSOR and/or a certified public accountant employed by LICENSOR, for the purpose of verifying LICENSEE's royalty statement or compliance in other respects with this license.
5.2 LICENSEE, within ninety (90) days after the close of each calendar quarter of each year, shall deliver to LICENSOR a true and accurate report, giving such particulars of the

business conducted by LICENSEE during the preceding three (3) month period under this license as are pertinent to a royalty accounting under this license. These shall include at least the following:

(a) Total number of units of LICENSED PRODUCTS sold by LICENSEE.
(b) Discounts allowed as defined in the definition of NET SALES in paragraph 1.5.
(c) Names and addresses of all sublicensees of LICENSEE.
(d) Total royalties due.

5.3 LICENSEE shall pay to LICENSOR the royalties due and payable under this Agreement quarterly, no later than ninety (90) days after each calendar quarter. If no royalties are due, it shall be so reported.

ARTICLE 6
TERMINATION

6.1 If LICENSEE shall become bankrupt or insolvent and/or if the business of LICENSEE shall be placed in the hands of a receiver, assignee or trustee for the benefit of creditors, whether by the voluntary act of LICENSEE or otherwise, LICENSEE shall immediately notify LICENSOR and LICENSOR shall thereupon have the right to terminate this Agreement by giving written notice to LICENSEE of such termination and specifying the effective date thereof, which shall be at least thirty (30) days after the date the notice is mailed by LICENSOR. Such notice shall be sent to LICENSEE by certified mail at an address designated as provided in Article 15 hereof, or to such other address as LICENSEE may designate from time to time in writing by notice to LICENSOR, and the rights, privileges, and license granted hereunder shall thereupon immediately terminate and neither party shall have any further rights, duties or obligations hereunder except as may have then accrued under this Agreement.

6.2 Upon any material breach or default of this Agreement by LICENSEE, LICENSOR shall have the right to terminate this Agreement and the rights and license granted hereunder by ninety (90) days notice by certified mail to LICENSEE. Such termination shall become effective unless LICENSEE has cured any such breach or default prior to the expiration of ninety (90) days from receipt of LICENSOR's notice of termination. In the event of such termination, the parties shall no longer have any rights, duties or obligations hereunder subsequent to the date of such termination, except as may have then accrued under this Agreement.

6.3 Upon termination of this Agreement for any reason, nothing herein shall be construed to release either party of any obligation which matured prior to the effective date of such termination, and LICENSEE and/or any sublicensee thereof may, after the effective date of such termination, sell all LICENSED PRODUCTS, and complete LICENSED PRODUCTS in the process of manufacture at the time of such termination and sell the same, provided that LICENSEE pays to LICENSOR the royalties thereon as set forth in Article 4 of this Agreement and the reports required by Article 5 hereof on LICENSED PRODUCTS.

ARTICLE 7
UNLICENSED ACTIVITY AND INFRINGEMENT

7.1 LICENSEE and LICENSOR shall promptly inform the other in writing of any license infringement by a third party and provide available evidence of infringement.

7.2 If within thirty (30) days after notification of alleged infringement LICENSOR has not been successful in persuading the alleged infringer to desist and is not diligently prosecuting an infringement action or if LICENSOR notifies LICENSEE of its intent not to bring action against the alleged infringer, LICENSEE or sublicensee with permission of LICENSOR may, but is not obligated to bring action at its own expense and may use the name of LICENSOR as party plaintiff.

7.3 In any suit involving the enforcement or defense of the licensed rights, the other party hereto agrees, at the request and expense of the party initiating such suit, to cooperate in all respects and to have its employees testify when requested and to make available relevant records, papers, information, samples, specimens and the like.

7.4 No settlement or consent judgment or other voluntary final disposition of an enforcement and/or defense suit initiated by either party to the Agreement may be entered into without the consent of the other which consent will not be unreasonably withheld.

7.5 In the event that a declaratory judgment action alleging invalidity or non-infringement of the licensed rights is brought against LICENSEE, LICENSOR reserves the right, within thirty (30) days after commencement of such action, to intervene and take over the sole defense to the action at its own expense.

7.6 If LICENSEE and/or any sublicensee thereof is required to pay a royalty or damages to another party resulting from a final judgment or settlement to which LICENSOR consents (such other party being hereinafter referred to as "THIRD PARTY LICENSORS") in order to make, have made, lease or sell a LICENSED PRODUCT, then and in the event, the royalty payable by LICENSEE to LICENSOR on such LICENSED PRODUCT shall be reduced by the amount of royalty that LICENSEE and/or sublicensee shall be required to pay to such THIRD PARTY LICENSOR. If LICENSEE avails itself of this provision, LICENSEE agrees to supply LICENSOR with proof of royalties paid to such THIRD PARTY LICENSOR.

7.7 The total cost of any infringement action commenced or defended solely by LICENSOR shall be borne by LICENSOR, and LICENSOR shall keep any recovery or damages derived therefrom.

7.8 The cost of any infringement action commenced or defended by LICENSEE shall be borne by the LICENSEE. The LICENSEE, however, may withhold royalties in any given calendar year and apply the same towards reimbursement of its expenses and any recovery of damages by LICENSEE from any such suit shall be applied first in satisfaction of any unreimbursed expenses and legal fees of LICENSEE relating to the suit or settlement thereof, and the balance of such recovery shall be paid proportionately to LICENSOR pursuant to Article 4.

ARTICLE 8
ASSIGNMENT

LICENSEE may assign or otherwise transfer this Agreement and the license granted hereby and the rights acquired by it hereunder so long as such assignment or transfer is accompanied by: (1) a sale or other transfer of LICENSEE's entire business or (2) sale or other transfer of that part of LICENSEE's business to which the license granted hereby relates, or (3) sale or transfer to one or more sublicensees. LICENSEE may assign or otherwise transfer this Agreement and the license granted hereby and the rights acquired by it hereunder if such assignment or transfer is accompanied by the transfer of the rights to manufacture and/or market a commercially viable product or products which embody the licensed TECHNOLOGY. LICENSEE shall give LICENSOR thirty (30) days prior written notice of such assignment and transfer. LICENSOR, however, shall not be deemed to have approved such assignment and transfer unless such assignee or transferee has agreed in writing to be bound by the terms and provisions of this Agreement in which event LICENSEE shall be released of liability hereunder. Upon such assignment or transfer of Agreement by such assignee or transferee, the term LICENSEE as used herein shall include such assignee or transferee.

ARTICLE 9
NON-USE OF NAMES

LICENSEE or its sublicensees shall not use the names of LICENSOR, nor any adaptation thereof in any advertising, promotional or sales literature without prior written consent obtained from LICENSOR in each case, except that LICENSEE may state that it is licensed by LICENSOR.

ARTICLE 10
UNITED STATES GOVERNMENT EXPORT CONTROL REGULATIONS

The Export Regulations of the U.S. Department of Commerce prohibit, except under a special validated license, the exportation from the United States of technical data relating to certain commodities (listed in the Regulations), unless the exporter has received certain written assurance from the foreign importer. In order to facilitate the exchange of technical information under this Agreement, therefore, LICENSEE hereby gives its assurance to LICENSOR that LICENSEE will not knowingly, unless prior authorization is obtained from the U.S. Office of Export Controls, re-export directly or indirectly any technical data received from LICENSOR under this Agreement and will not export directly the LICENSED PRODUCT or such technical data to any restricted country.

LICENSOR neither represents that a license is or is not required nor that, if required, it will be issued by the U.S. Department of Commerce.

ARTICLE 11
HOLD HARMLESS

Except for legal actions included within Article VII of this Agreement, LICENSEE shall defend or settle, at its own expense, any claim, action, suit or legal proceedings which may be brought against LICENSOR by reason of the manufacture or distribution of LICENSED PRODUCT by LICENSEE and will indemnify and hold LICENSOR harmless from and against all damages and costs adjudged or decreed against and actually paid by LICENSOR in any such claim, action, suit or legal proceeding in accordance with a final decree of final judgment rendered by a Court of Competent Jurisdiction in a decision, unappealed or unappealable, or any costs actually paid by LICENSOR in connection with any settlement of any such claim, action, suit or legal proceeding; provided, however, that LICENSOR shall have given LICENSEE prompt written notice of such claim, action, suit or legal proceeding and shall permit LICENSEE, by counsel of its own choosing, to defend or settle same, and provided further that LICENSOR shall not have settled such claim, action, suit or legal proceeding without the consent of LICENSEE, which consent shall not be unreasonably withheld. LICENSEE agrees to name LICENSOR as named insured on any product liability insurance obtained by LICENSEE.

ARTICLE 12
BENEFITS OF LITIGATION, EXPIRATION OR ABANDONMENT

In case any patent within the patent rights granted hereunder expires or is abandoned, or is declared invalid or otherwise construed by a court of last resort or by a lower court from whose decree no appeal is taken or certiorari granted within the period allowed therefor, then the LICENSEE shall have the right to terminate the Agreement in accordance with Article 6.2.

ARTICLE 13
PATENT COSTS

13.1 The LICENSEE shall reimburse LICENSOR for patent costs up to the amount of __________ payable __________ from the date of execution of this Agreement,

13.2 The LICENSEE shall reimburse LICENSOR for the annual maintenance fees associated with all foreign patents issued in connection with licensed TECHNOLOGY.

ARTICLE 14
MISCELLANEOUS PROVISIONS

14.1 This Agreement shall be construed, governed, interpreted and applied in accordance with the laws of the State of New York, U.S.A., except that questions affecting the construction and effect of any patent shall be determined by the law of the country in which the patent was granted.

14.2 The parties hereto acknowledge that this instrument sets forth the entire Agreement and understanding of the parties hereto as to the subject matter hereof, and shall not be subject to any change or modification except by the execution of a written instrument signed to by the parties hereto.

14.3 The provisions of this Agreement are severable, and in the event that any provisions of this Agreement are determined to be invalid or unenforceable under any controlling body of law, such invalidity or unenforceability shall not in any way affect the validity or enforceability of the remaining provisions hereof.

ARTICLE 15
PAYMENTS, NOTICE AND OTHER COMMUNICATIONS

15.1 Any payment, notice or other communication pursuant to this license shall be sufficiently made or given on the date of mailing if sent to such party by certified air mail, postage prepaid, addressed to it at its address below or as it shall designate by written notice given to the other party:
In the case of LICENSOR: ______________
In the case of LICENSEE: ______________

15.2 Wherever the consent of LICENSOR is required under this Agreement, and LICENSEE has given prior written notice and LICENSEE raises no objection in writing within the required period of time after the giving of such notice, or thirty (30) days in the event no required period is stated, then LICENSOR shall be deemed to have approved any action stated in the notice.

IN WITNESS WHEREOF, the parties hereto have hereunto set their hands and seals and duly executed this License Agreement the date first above written.

LICENSOR:

By: ____________________________ / Date: ___________

Title: ____

LICENSEE:

By: ____________________________ / Date: ___________

Title: ____

BIBLIOGRAPHY

Asimov, Issac. *Asimov's New Guide to Science.* New York: Basic Books, Inc., Publishers, 1984.

Association of University Technology Managers, Inc. Survey Conducted by Diane C. Hoffman, Inc.: *The AUTM Licensing Survey, Executive Summary and Selected Data, Fiscal Years 1993, 1992, and 1991.* Connecticut: Association of University Technology Managers, Inc., 1994.

Atkinson, Stephen H. "University-Affiliated Venture Capital Funds," in *Health Affairs* I (Summer 1994).

Brandon, Bobbi A. "Deposit Requirements for Patent Purposes," in *Biotechnology Patent Conference Workbook.* Maryland: American Type Culture Collection, April 1989.

Brody, Richard J. *Effective Partnering: A Report to Congress on Federal Technology Partnerships.* U.S. Department of Commerce, Office of Technology Policy, April 1996.

Bugbee, Bruce W. *Genesis of American Patent and Copyright Law.* Washington, DC.: Public Affairs Press, 1967.

Bygrave, William D. and Jeffry A. Timmons. *Venture Capital at the Crossroads.* Boston, Massachusetts: Harvard Business School Press, 1992.

Cardwell, D. S. L. *Technology, Science and History.* London: Heinemann, 1972.

Cooper, R. G. and U. de Bretani. "Criteria for Screening New Industrial Products," in *Industrial Marketing Management.* Vol. 13 (1984).

Degnan, Lauren A. "Does U.S. Patent Law Comply with TRIPPS Article 3 and 27 with Respect to Treatment of Inventive Activity," in *Journal of the Patent and Trademark Office Society.* Vol. 78, No. 2 (February 1996).

Dilworth, Peter. "Some Suggestions for Maximizing the Benefits of the Provisional Application," in *Journal of the Patent and Trademark Office Society*. Vol. 78, No. 4 (April 1996).

Dorland, Gilbert N. and John Van Der Wal. *The Business Idea*. New York: Van Nostrand Reinhold Co., 1978.

Einhorn, Alan H. *Overview of Federal Law on Technology Transfer* (Manuscript distributed at 1994 Conference on Commercializing Biomedical Technologies).

Ergas, Henry. "Does Technology Matter," in *Technology and Global Industry*. Ed. by Bruce R. Guile and Harvey Brooks. Washington, D.C.: National Academy Press, 1987.

Farnsworth, Allan E. An *Introduction to the Legal System of the United States*. New York: Oceana Publications, 1983.

Federal Register. Vol. 59., No. 23, pp. 33308-33311.

Finnegan, Marcus B. and Herbert H. Mintz. "Determination of a Reasonable Royalty in License Negotiations, " in *Licensing Law and Business Report*. Vol. 1, No. 2 (June-July 1978).

Fox, Harold J. *Monopolies and Patents: A Study of the History and Future of the Patent Monopoly*. Toronto: The University of Toronto Press, 1947.

Ginsburg, Douglas H. "Antitrust, Uncertainty, and Technological Innovation," in *The Antitrust Bulletin* (Winter 1979).

Goldner, Howard J. "Analytical Instrumentation," in *R&D Magazine* (September 28, 1992).

Goodwin, John R. *Business Law Principles, Documents and Cases*. Illinois: Richard D. Irwin, Inc., 1976.

Hall, Christopher H. "Renting Ideas," in *The Journal of Business*, Vol. 64, No. 1 (January 1991).

Hindle, Brooke and Steven Lubar. *Engines of Change: The American Industrial Revolution, 1790 - 1860*. Washington, D.C.: Smithsonian Institution Press, 1986.

Holum, John R. *Introduction to Organic and Biological Chemistry*. New York: John Wiley & Sons, Inc., 1969.

Jacobs, Alan J., ed., updated by Elizabeth Hanellin. European Community Patent Convention. *Patents Throughout the World* (New York: Clark, Boardman, Callaghan, 1994).

Kalas, John W. *The Grant System*. New York: State University of New York Press, 1987.

Kuznets, Simon. *Modern Economic Growth: Rate, Structure, and Spread*. New Haven, Connecticut.: Yale University Press, 1972.

Leikind, Morris C. and Wyndham Miles. *The Nature of Science and Technology, in Science and Technology: Vital Resources*. Ed. Ralph Sanders. Maryland: Lomond Systems, Inc., 1975.

Mackay, Harvey. "Mackay on Business: Sooner or later, the day will come when we have to learn to be losers" in *Times Union*. Sunday, May 21, 1995.

Mathur, Krishna D. and Sanders, Ralph. "Science and the Federal Government. Science and the Federal Government," in *Science and Technology: Vital National Resources*. Ed. Ralph Sanders. Maryland: Lomond Systems, Inc., 1975.

Maynard, John T. *Understanding Chemical Patents: A guide for the inventor*. Washington, D.C.: American Chemical Society, 1978.

McCartt, Anne Taylor. "The Application of Social Judgment to Library Faculty Tenure Decisions," *College and Research Libraries*, September (1983).

McEachern, A. W. "Two Simple Versions of Multivariate Utility Analysis," in *Decision Making in the Public Sector*. Ed. Lloyd G. Nigro. New York: Marcel Decker, Inc., 1984.

Miller, Thornton F. *Federal Common Law in Historic U.S. Court Cases 1690-1990: and Encyclopedia*. Ed. John W. Johnson. New York: Garland Publishing, 1992.

Morehead, John W. *Finding and Licensing New Products & Technology from the U.S.A.* Elk Grove Village, Illinois: Technology Search International, Inc., 1982.

Muir, Albert E. "Interindustry Analysis of the Impact of Federal Support for Academic Science on the Economy of New York State," *Research in Higher Education*, Vol. 18, No. 2 (1983).

—. "Managing Inventions Marketing," in *les Nouvelles (Journal of the Licensing Executives Society)*, Vol. XXV, No, 4 (1990).

—. "Rationalizing Royalties," in les *Nouvelles (Journal of the Licensing Executives Society)*. Vol. XXI, No. 2 (1986).

—. "Technology Transfer Office Performance Index," in *Journal of the Association of University Technology Managers*, Vol. V (1993).

Narin, Francis and Dominic Olivastro. "Status Report: Linkage Between Technology and Science," in *Research Policy* 21 (1992).

Narin, Francis, Elliott Norma and Ross Perry. "Patents as Indicators of Corporate Strength," in *Research Policy* 16 (1987).

National Science Foundation. *National Patterns of R&D Resources*. National Science Foundation, 1992.

National Science Board. *Science Indicators, The 1985 Report*. National Science Foundation, 1985.

Neale, A. D. *The Antitrust Laws of the U.S.A. 2nd Edition*. London: Cambridge University Press, 1970.

Ninety-third Congress, 1st Session, House Document. *Historical Statistics of the United States, Colonial Times to 1970*. U.S. Department of Commerce, Bureau of the Census.

Noble, David E. *America by Design - Science, Technology and the Rise of Capitalism*. New York: Alfred A. Knopf, 1977.

Orenbuch, Louis. "Trade Secret and Patent Laws," in *Journal of the Patent Office Society*. October 1970, Vol. 52, No. 10.

Preston, John T. "Key Problems in Commercializing Technology in the U.S." *Presented as Testimony Before the Energy Subcommittee of the House Space Science and Technology Committee* (March 1993). Copies of material distributed for a 1995 presentation at the Rensselaer Polytechnic Institute, New York.

The Research Foundation of State University of New York. *Strategic Plan*, 1996.

Rosenberg, Peter D. *Patent Law Basics*. New York: Clark, Boardman, Collagham, 1992.

Rostoker, Michael D. *A Survey of Corporate Licensing, Patent, Trade Secret, Know-How*. Connecticut: The Franklin Pierce Law Center, 1984.

Rostow, W. W. *The Process of Economic Growth*. New York: W. W. Norton & Co., Inc., 1962.

Saliwanchik, Roman. *Legal Protection for Microbiological and Genetic Engineering Inventions*. Massachusetts: Addison-Wesley Publishing Co., 1982.

Scherer, F. M. *Industrial Market Structure and Economic Performance*. Chicago: Rand McNally College Publishing Co., 1970.

Schmookler, Jacob. *Invention and Economic Growth*. Cambridge, Massachusetts: Harvard University Press. 1966.

Schnooker, Alan N. *Negotiate to Win - Gaining the Psychological Edge*. New Jersey: Prentice Hall, 1989.

Small Business Development Center and The Research Foundation of State University of New York. *Workshop Concerning a Firm's Ability to Launch a New Product*, held at the Rockefeller College in Albany, New York in 1989.

Speers, Mark, Paula Ness Speers, and Lisa Pisano. "Choosing the Right Licensee(s)," in *AUTM Technology Transfer Practice Manual*. Connecticut: Association of University Technology Managers, 1993.

State University of New York. *Inventions and Patent Policy.*

Technology Touchstone. *NTTC/Knowledge Express Enter Partnership*. Vol. 3 No. 1 (March 1995).

Thackray, Arnold. "University-Industry Connections and Chemical Research: An Historical Perspective," in *National Science Foundation: Selected Studies.*

Udell, G. G. and K. B. Baker. *PIES-11 Manual for Innovation Evaluation*. Madison, Wisconsin: University of Wisconsin - Extension, 1984.

Van Horn, Charles E. "Effects of GATT and NAFTA on PTO Practice," in *Journal of the Patent and Trademark Office Society*. Vol. 77, No. 3 (March 1995).

U.S. Government. *Technology Transfer Legislative History.* TechTRANSIT Home Page. Internet, http://www.dtic.dla.mil/...Legislative_History.html. (December 7, 1996).

U.S. Department of Commerce. *Statistical Abstracts of the United States 1993.*

U.S. Department of Justice. *Justice Department Release of Draft Intellectual Property Guidelines for Public Comment.* August 8, 1994.

U.S. Patent and Trademark Office. *Highlights in Patenting Activity*. U.S. Department of Commerce.

—. "Legal Analysis to Support Proposed Examination Guidelines for Computer-Implemented Inventions," in *Official Gazette*. U.S. Department of Commerce, November 7, 1995.

—. *Patent Laws*. U.S. Department of Commerce. Washington, D.C.: Government Printing Office, 1976.

—. Technology Assessment and Forecast Program. *Patenting by Organization*. U.S. Department of Commerce, 1994.

Weiner, Charles. "Universities, Professors, and Patents: A Continuing Controversy," in *Technology Review*. February-March 1986.

Wilson, Mitchell. *American Science and Invention: A Pictorial History*. New York: Bonanza Books, 1960.

—. *American Science and Invention: A Pictorial History*. New York: Bonanza Books, 1974.

World Book Encyclopedia. Vol. 11. Chicago, London: Field Enterprise Educational Corp., 1974.

—. Vol. 15. Chicago: Field Enterprise Educational Corporation, 1974.

INDEX

ABOUT THE AUTHOR

Albert E. Muir. The author has seventeen years experience in marketing, licensing and patenting inventions. As Licensing Coordinator, he manages technology transfer for The Research Foundation of State University of New York (Central Office). Articles by Dr. Muir appear in the *Journal of the Association of University Technology Managers (AUTM), les Nouvelles* (Journal of the Licensing Executives Society) and *Research in Higher Education*. Dr. Muir received his undergraduate degree from Ithaca College, and his doctorate degree from the University at Albany, State University of New York.